Genetics For A New Human Ecology

Michael Breeden

Copyright © 2017 Michael Breeden

All rights reserved.

ISBN: **1544900996**

ISBN-13: **978-1544900995**

DEDICATION

Dedicated with great gratitude to all those with patience and to all those that helped.

CONTENTS

1	Preface	1
2	Introduction	9
3	Why – Genetic Load	17
4	How – Artificial Selection	31
5	Morality of Genetic Husbandry	37
6	What – Health, Beauty and Brains	47
7	The Three Levels of Artificial Selection	55
8	Morality of Humanity	101
9	About Racism	137
10	Wrap Up	149

Humanity is like the crew of a ship that has left their home and must find a new place to live. Unfortunately, their ship is sinking because of many growing leaks, they are running out of supplies, they are sick and getting sicker and they do not know where they are going. Plus they are crowded together and fighting among themselves. Storms could come at any time that would make it worse or finally sink them. This book is to tell them how to get healthy again, how to fix the leaks, make a peace that can work for all, survive the storms, and where they can find land. It describes the new land as a wondrous place of great riches where they can live, grow, and raise families.

Humans are in a dangerous time of change. We are in between. We have left the tribal world that we came from and must find a new place to live. The very ecology that is our life support system is badly damaged. The dangers are closer than they seem, but the potentials are as great as human aspiration.

1. Preface

This is Book One of the Transition Series that describes how humans can adapt genetically and strategically to a new ecology to replace the tribal ecology we left when we started cities and farming. The dangers are great and near, but the potentials are everything humans have aspired to. This is how the series of books is laid out to tell this story.

Genetics For A New Human Ecology - How humans can adapt genetically to a changing world. It focuses on the incredible danger we face and the potentials that come from it.

Strategy For A New Human Ecology - How humans can adapt strategically. This not just about some future. This is about current problems such as tribalism and automation. It is how to make human civilization work. (In process).

Power - The Darwinian instinct for dominance

drives a mindless quest for power that has no off switch. Eventually, it competes with and endangers civilization. We need to deal with that danger strategically and philosophically. (In process).

When Barbara Explained Genius - This is a description of intelligence and how it works. It is the behavior that got us here and it will be the most important behavior in the future.

Transition To A New Human Ecology - The whole ball of wax. Good luck getting through that monster, but it contains many ideas that may not be in the shorter stories.

The world is rapidly changing. That should be pretty obvious to anyone. What is changing and what it means is another story. Everyone has a perspective on that, but it is like the story of the blind men describing the elephant. They can only know about a part of it that they experience. What we commonly see is different descriptions of the changes occurring in terms of economics, politics, technology, etc. Or we hear of problems like over population, war, environmental disasters and that kind of thing. But we mostly see isolated views. What if there was a way to organize all these different points of view so we could see the whole picture rather than just isolated parts? Is there a science for organizing all these changes and problems all together at once? There is. It is ecology. It describes a species and the world it lives in. That might be an amoeba in a pond or an elephant in a forest, but for humans it is much more complicated. We are a complicated species living in a large, complicated world. Ecology can organize all these different views people have to show a larger, detailed, connected picture of humanity and the human world. That is what this is for so that it can lead to a broad, integrated understanding of humanity, our needs, the problems we face, how to solve them and the rich potentials in our future.

Back when I was in high school biology, my teacher mentioned that there were diseases that we never encountered and knew nothing about (maybe something like COVID-19, maybe something worse). Since I liked biology, it got me to wonder about how using modern medicine and vaccines to reduce disease and other medical problems would effect humans. The effect is big. Thinking about it, the issues seemed important and far-reaching because of how many people normally died from diseases. Also, diseases constantly mutate so the problem never ends. It was fascinating so I kept exploring it, assembling the data, organizing it, and trying to figure out what it meant. Eventually, it all became a biological description of humans, using the science of ecology for organizing all the information, which is the standard tool used by science for describing a species. It revealed a lot. The more I looked and the more I understood about humans, the more I was amazed at the magnitude of the changes we were experiencing in just about every aspect of human life, habit, strategy, and seemingly our very existence. It is what we usually call society, religion, family, resources, economics, politics, and every other institution humans have been developing over and over again for the past thousands of years since we started the cities. Also though, our nature was changing due to the genetic coming together of peoples and harsh genetic selection.

Ecology is a great tool for organizing the description of all the changes and their effects. It is a very good tool for managing what is a complicated problem and a great deal of data. It shows connections and relations that would not be seen otherwise. At the same time using ecology as the tool seemed exceptionally effective because it appeared that all the dangers and changes I was seeing could be attributed to fundamental changes in human ecology occurring because we were leaving one ecology and developing another. We were leaving the niche of the so-called hunter-gatherer that we grew up in. We started farming and creating cities, but where we were going was not so clear though. All species do need a fairly stable ecology to survive in. It is basically their life support system. Being between ecologies is a dangerous place to be. The creation of cities and farms was the start of a new human ecology. There was a lot of data available to examine for what that might mean, including lots of information to work with from biologists, anthropologists, literature, even religion, and lore, as well as many thousands of years of history. More and more, genetic information has become available as well. I spent many years assembling and organizing the data, trying to see what it showed. I was hoping to find suitable solutions to the dangers I saw and how to take advantage of the possibilities of that new ecology for both human survival and growth. I wanted to make some kind of theoretical concept of how humans could again create a relatively long-term stable ecology where the current dangers of being between ecologies were behind us. It would be a world where changes were not occurring at the current mad pace, and where humanity could make their own destiny instead of so often just reacting to events like animals have to. While the future is largely impossible to predict, it can be described in terms of ecology. We do know what that requires, and this can show what is important or even parts that might be missing from our views. It is though, not a given that we can make it to this new ecology. Our ancestors have worked for this for thousands of years without knowing what they were even working for. This can explain what they were working for, this new world they wanted to build. Having a goal is a great help when you have started a journey like humanity is on. This can explain why we should continue to work to build and preserve this new world for humanity. It is a good world. It will be a new ecology where humans can survive and grow. It is our only chance to be more than animals. I did not start out trying to describe a new human ecology where humans could survive. I just was curious about what changes widespread vaccinations would cause but the far larger picture that I found amazed me and gave me a much larger understanding of what is happening now during these larger changes. I hope you like this story and that it explains a lot to you. The changes are pretty obvious. This is what the changes mean in terms of human genetics. Some strategy is considered but that is mostly meant to be in another book.

* * *

This book is basically about biology but it is not meant to be very advanced biology. It is just biology that is examined very closely, with a fair amount of logic applied to it. Since it is fairly basic, the biology is supposed to be familiar, but the close examination and reasoning is supposed to show new things. This offers an understanding unlike any other.

At the same time, this is about how humans can survive and grow so it is actually a morality play. It describes an existential danger humans face and how we can overcome it. It describes how overcoming that can also lead humanity to a bright future of great potential and the achievement of great aspirations. Yes, I hope you find it an interesting read that gives you new ideas, but its real purpose is as a solution for when that great danger threatens us and people want to know what to do. It should describe how to solve the problem and lead to a bright future for humanity.

2. Introduction

This book is about human genetics in the context that humans are between two ecologies. The so-called hunter-gatherer tribal ecology we left that was replaced by farms and cities, and whatever ecology will follow that. This book focuses on the genetic parts of how we can adapt to that new ecology, rather than the strategic parts which will come later. It turns out though that this discussion must primarily focus around one thing because it is an extremely dangerous genetic problem that is already starting to effect us. Some of the changes we have already made, such as parents being older, families being smaller and medicine, have changed the balance between the natural mutation rate and the rate of natural selection. The weak and sick are not being removed from the gene pool by nature like they used to be. This is going to lead to a disaster because generation by generation we are just going to naturally accumulate genes that have become broken by (de novo) mutations. Unless we can replace natural selection some, the birth defect rate will rapidly rise and keep rising until we lose our civilization and natural selection returns to its normal level. Nicely, there is an economical and ethical way to solve this problem, that will allow everyone to have healthy families. It will even allow us to become better adapted to the ongoing changes we face and to a good future where humanity can live and grow long term. This starts an examination of the problem of this imbalance leading to genetic defects and how we might solve it by husbanding our genes. It also offers the potentials to solve many other genetic and strategic problems, including racism, that we will need to solve for humans to have a bright future. It is not a future for machines as some like to tell the story. It is a future for humans and not just any humans. This is for humans that are far more than just animals. No, it is not about "superhumans". I do not know what those are anyway. It is about humans like us but a good deal healthier, more beautiful, and a bit more intelligent. This is a story of how we can become physically and mentally strong enough to survive in a new world that we must create, with the wisdom to know how and why. It is about making a world we want to live in and where we will have time to decide what we do want to become. The potentials are amazing.

There is a second lesson to this story. Nature has incredible strategic limitations that human thought can easily overcome. Humans can come up with much better solutions than nature can. Nature proceeds by natural selection, a simple, blunt, and brutal instrument driven by blind competition that can only make very simple, short-term decisions. Humans can make longer-term, more complicated decisions that will allow far better outcomes than nature can. This is true both genetically and strategically. We need to. The strategic aspect of this will be explored more in the next book about strategies of survival but is very clearly illustrated in terms of genetics by the limitations of natural selection compared to the potentials of human strategies of genetic husbandry discussed in this book. What do you respect about yourself that you would like your children to inherit? What would you prefer that they did not inherit? The way nature works, is like flipping a coin. The chances of any outcome are basically 50-50. If we use human strategies to husband our genes, the odds change. It will be like rolling a dice where five (or more) sides of the dice have the desired outcome and only one side is not what you want. It is like flipping a coin to make your decisions compared to making carefully thought-out decisions. Human thought can produce far better outcomes than the "coin tossing" that is all nature can do. The greatest wealth humans can have is in their genes and we can increase that wealth incredibly using human strategies to husband our genes. We will need to in order to survive as more than animals. Humans are pretty amazing. We can become far more than we are. The important lesson is larger though. Nature gives survival strategies to all organisms. Just as the genetic strategy of natural selection is not a very good one that has great drawbacks and can only work slowly over vast amounts of time, so too some of our natural behavioral strategies are really poor and they tend to have no off switch. Just as we can use and need to use human strategies to husband our genes to get a much better outcome than the strategies of nature, our behavioral strategies are the same. Many instincts and behaviors we have that are a result of Darwinian forces are not very good strategies. Some will actually be dangerous to us, such as our tendency towards dominance and violence. Luckily, we have other instincts that will work much better, such as ones that allow cooperation. We just need an understanding of what instincts we have available, which ones to use, why, and at what times. We need to make decisions about what strategies to use. That is the focus of the Strategy book. The focus of this book is when and how and why to use human-created genetic strategies instead of the strategies of nature.

*   *   *

Genetics is one of the most important current topics in any case, but it must be considered in a larger context. The human world is not only changing rapidly but also changing in deeply fundamental ways. All of these changes present dangers, challenges, and opportunities, but none present more danger and potential than the changes effecting our genes. This discussion is about genetics in that rapidly changing world. In biological terms, it would be said that humans are undergoing the greatest change in ecology that any species has ever survived, if we succeed. So this is about genetics in the context of how humans can create and adapt to a new relatively stable ecology where we can survive and develop long-term. We need to do that because the ecology that we grew up in and are most adapted to, is gone. The so-called hunter-gatherer ecology has been replaced by the farms, towns, and cities of civilization. Tribes have been replaced by larger societies. Since we left that way of life, we have been in transient ecologies, which is a dangerous place for a species to be, because a species without a stable ecology is called... "extinct". So we need to create and adapt to this new ecology. It is going to be deeply different because it will be an ecology that is not created by nature so much as it will be created by human strategic and genetic adaptation. For now, that description does include nature as most ecological descriptions do, but human ecology is also about buildings, factories, machines, power production, communication, etc. and it will become more so as time goes on. Still, it is an ecology. It is still our life support system. The unique thing about humans is our ability for strategic adaptations. In this case it is how we can adapt our genetic strategy. There is far more to human ecology as well and I think you would be surprised how much it reveals about the problems we face now. Again though, this book is about the genetic issues.

Humanity is at the dawn of a new age as defined by many things. One of the most important will be that from now on we will need to take responsibility for our genetic health instead of relying on Nature to take care of it. We have no other choice.

Heredity has been called the Second Forbidden Question In Science. The problem is that the science of heredity has been used in the past to rationalize racism and to justify what were race wars that were some of the darkest events in human history. Human genetics must be examined though, both because of the great wealth it represents and because of the great danger that comes from changes we have already made effecting our genes. At the same time, genetic technology is exploding and must be developed in a framework of moral understanding and human benefit rather than just exploitation and profit-taking. Genetic knowledge can lead to great monetary wealth, but this is even more about human wealth. This book clearly states its basic moral context, but this is mostly focused around the consequences of just one aspect of human genetics, a danger we must address soon. A larger moral context related to genetics is detailed in the strategy book, but a main point contained here is that this work is written so that it inherently cannot be used to support racism. It clearly describes the danger of racism to any human future. Human progress has always been by the coming together of peoples. Trying to stop that will close off our future.

3. Why – Genetic Load

Generally, I refer to the new ecology that humans are developing as "Civilization" which is an unusual description for an ecology, but then it is an unusual ecology. In a tribal ecology, humans usually dealt with people that were like themselves. In civilization, we commonly deal with people that are more different. Civilization is the new form of our society, as opposed to tribal. In terms of resource strategy, tribal ecology is "hunter-gatherer-scavenger" of wild crops and game as compared to the "technical" strategies of civilization that uses resources unavailable to animals.

Genetically adapting to this new ecology is going to be a long challenging process, especially because it is a fairly novel ecology we need to adapt to. It is going to take time, but that is called life. It is called survival and we can do it. Our genes offer great potential. Unfortunately though, there is another genetic issue that is much closer at hand that we need to deal with. That issue is that the changes we have already made to our habits and ecology mean that we have increased the net mutation rate while reducing the natural selection rate. This is going to lead to a very dangerous genetic load that we cannot survive with.

Genes are pretty amazing. Genes are the blueprint of life. Genes not only describe all life from a fungus to an elephant, but they also describe how to get from a single cell to an elephant. Genes even have instructions for how to create a mechanism to reproduce themselves, that is us. They have instructions for how to mix and rearrange themselves with other genes to increase their adaptability. Your genes are copied repeatedly all through life as new cells replace old cells. The thing is that while there are mechanisms to ensure that the copy processes go smoothly and accurately, it is not always perfect. Sometimes random changes occur then and during the recombination process related to sex cells. We call these mutations. Usually, they are just a simple exchange of a couple of base pairs on a gene and are not such a big problem. Some have essentially no effect (up to even 15% of the single base pair switches might not change things). Typically, an individual may have a few thousand of these single point mutations where one base pair is changed. They usually are not much trouble unless two of them occur at the same location or perhaps under a stress like disease. At the same time, while mutations may not be a problem, other times they can be and they certainly are not likely to be an improvement or even good. They are random changes to the genetic sequences that describe you. Mutations are almost never good. Your genes are the creation and refinement of many thousands and even millions of years of natural selection and evolution to create the most fit species and most fit individuals, in order to enable and ensure survival. That is what life is about. That is what nature and evolution are about, survival. Random changes are very unlikely to be an improvement and are far more likely to reduce the functionality of any gene, reducing the fitness of the individual. Natural selection is the mechanism that removes individuals with mutations that reduced their fitness so that the genes of the species remain fit and the species survives. Natural selection is a brutal and very blunt instrument of control, but it is how nature works and is how species survive.

Not all mutations are bad. Occasionally, mutations

are actually good. They are the raw material from which evolution creates species and allow a species to adapt to its environment or to changes in its environment, but still, being a random thing, mutations are very rarely an improvement.

We usually think of mutations as little genetic changes in the DNA base pairs, but often they are not so little. Sometimes these copy errors may be big changes in the genetic sequence. Parts of the genes may get scrambled, be duplicated or even fail to be copied. There may be a failure to copy an entire chromosome or there may be multiple copies of chromosomes created. Small or large, changes like these are very rarely good and may cripple a person or cause lifelong health problems, if the zygote even survives to birth. They can then be quite fatal individually in terms of natural selection or fatal in evolutionary terms, meaning they prevent reproduction. Mutations are a very general effect though and may just cause a weakness in the individual or cause problems under certain conditions. These may get passed on to the next generation, though commonly the result just seems to be infertility. These genetic errors, called mutations, will accumulate and add up as what is called "Genetic Load". It may even take a while, but eventually, the individual will "fail" the test of fitness that is each generation and these "bad" mutations will be removed by Natural Selection.

The mutation rate of a species is fairly constant and in balance with the natural selection rate. The problem is that since we left our last ecology, we have made changes that result in a net increase in the mutation rate and a very significant decrease in the natural selection rate. This will lead to a genetic load of broken genes that will reduce the general fitness of a species, us humans, and its individuals. Our current situation of increased mutations and reduced selection cannot be maintained and something will change.

What makes these changes so problematic is that we have already changed some important habits. It used to be that typically humans started reproducing in their mid-teens. Now we have changed it to where it is considered ideal for a woman to start a family in her early thirties when she is more mature and better positioned in life. Unfortunately, there are a lot more copy errors created in the sex cells as we age. This is especially true for women and for a woman of age 35, it is calculated that half of her eggs will have significant genetic problems. This effects men as well, but not as much. The problem relates to reduced oxygen levels around the woman's ovaries. By age 45 it is calculated that 95% of her eggs have significant abnormalities in their genes. To reduce this problem, there is the possibility of freezing eggs when a woman is young and fewer mutations have occurred, for later use when she is ready to have a family. Perhaps a tissue sample could be saved from youth and manipulated to produce sex cells later. Unfortunately, that will not effect what happens at recombination. The basic problem will still occur and it is not the only problem or even the main part of the problem. At the same time that copy errors have been increasing, we have been greatly reducing natural selection. We call it human progress. It is impossible to completely remove natural selection, but we have greatly reduced it and in particularly important ways. Partly it is advances in medicine, including vaccines. It is also better nutrition which used to contribute to a lot of mortality in infants and children. It used to be that in many societies, children were not named until after the first year, (or as they looked at it, until after the first winter) because the infant mortality rate was so high. That does not even include the natural selection rate before birth that is from known and unknown miscarriages. In history, typically no more than 50% of children born survived to reproduce. That is a pretty high selection rate.

More recent research says that men are effected by epigenetic load as genes in their sperm accumulate methylation that can inactivates genes. This is early research, but it suggests another cause for the same issue of genetic problems caused by older parents. (Alexander Suvorov, University of Massachusetts Amherst.) Tissue culture seems more and more attractive for creating sex cells. That definitely does not mean cloning, as explained below, that skips a step in evolution and would defeat the great potential benefits of recombination.

There are other things humans have done to reduce natural selection. This ecology is usually just a safer world than the tribal ecology was. We are not so often chased by lions and tigers and bears... and snakes. We are also less subject to homicide which used to be rather common.

The first topic I studied was disease and it is what

made me realize there was this huge reduction in natural selection. The effect of disease is an interesting and uniquely general selective effect. Humans have various mechanisms for defeating diseases, some are fairly targeted with fairly specific genetic foundations. Some immune responses are fairly general though, including fever. Many times, the body just heats up to try to kill an invader, which it may do or it may fail and kill the human host first. That lethal failure can be caused by the failure of any essential bodily system. So if there is a weakness, perhaps even created by one single point mutation, the body fails and natural selection occurs. That makes disease a uniquely general selective effect. It selects against weak links in the person's entire genome. We have seen this effect with the recent COVID virus. It hits people at their pre-existing health problems and may be fatal due to that single weakness.

There is another huge change that humans have made in terms of ecology though and it is the biggest change that has reduced natural selection. Reproduction is one of the two foundations of ecology and we have changed our reproductive strategies from quantity to quality. We used to basically have families as big as possible and hope that some of the healthiest, fittest children would survive to reach maturity and start their own families. Consider that statistically speaking, half of the children born are more fit than their parents and half are less fit than their parents. Really it is a bit less than half that are as fit as their parents, because of the ongoing mutation rate. Nature never intended the incredibly high survival rate that humans now have. Nature does not care. If you are familiar with big families, you may have observed this. Some children are just stronger, some are weaker. Some children of a family are amazing, some just do not seem to be in the game. Smaller families reduce the opportunity for natural selection to act.

There is another form of selection that used to effect humans to different degrees at different times, that could actually be called artificial selection and could be very important. One part of that was called "rites of passage", something common to almost all tribal and even later societies. If during this, the individual failed to show that they were able to fulfill their role in the tribal society, the consequences might be many things including death or banishment, but the point was that they were excluded from the society and that meant that they were excluded from having children. Be assured that the Greeks were not the only ones that used exposure to weed out weak offspring. Many warrior tribes executed the unfit or cowards. In many ancient societies were functions like the censors of Rome. They could judge a family worthy in any terms including moral fitness and it had survival consequences. A lot of history is a bit blurry, but the ruling classes did descend from herders and they were acutely aware of heredity. How much it was done is debatable, but it is certain that the ruling classes intentionally bred themselves. With the development of nations though, this changed. Social testing of this sort might very well be more oriented towards loyalty than fitness. Explicit selection pressure by the society has sort of gone away with the rise of democracy. Then again, that is what status is, which group you have reproductive access to. Status is a behavior older than humans and so shows a selective effect different than most that make up natural selection. It gets a little blurry there also. Status in humans is conscious and instinctive, considered decisions about fitness, by an individual about potential mates. We try to "marry" the fittest mate. It should probably still be thought of as natural selection though because of its strong genetic basis. Pretty much all species with any kind of intelligence have the capability to judge the fitness of other members of their species. In humans, "beauty" is one of the commonest indicators of health and fitness, though it can also contain elements of deception. In any case, these were important forms of genetic selection that are generally no longer practiced or are greatly changed. Often, status has changed because of material wealth replacing genetic wealth, though the instincts related to status are still extremely important and powerful.

Just numerically, with smaller families, we have greatly reduced the numbers that natural selection can act upon. Humans have always had a much longer developmental period compared to other animals. Now, as the world has gotten more complex, it takes even longer and takes more resources to raise a child to maturity and independence to where they can start their own family. We have much smaller families now and devote far more resources to each child. This is a very significant change and requires a lower mortality rate, as the demands of raising children in the modern world have risen greatly. Having a smaller family forces us to fight natural selection which leads to this imbalance between mutation and natural selection, leading to an increasing load of broken genes.

If this continues, at some point women are going to be told that between one-third to one-half of their children will be born with significant genetic defects including infertility. Researchers using inexpensive genetic sequencing are experimentally detecting this problem now. They have studied at-risk groups such as mental retardation, the most expensive disability, and premature babies that can have health problems all through life. They found that between 33% and 66% of the effects were from genes that they are calling "de novo" mutations, which is Latin for "fresh" because neither parent has these genetic sequences. They are not simply fresh though, they are copy errors and they are broken genes. Further recent research is showing that this is the cause of many other health problems, including many cases of autism, mental illness, cancers, and the endless list of human frailties. Many of these small mutations do not result in obvious disabilities or fatality, but these broken genes, that I used to call non-integral genes, do accumulate generation by generation. Normally natural selection would be removing them, but now it may not. A species cannot survive without a healthy gene pool any more than an individual can, and at the same time, we need to adapt to the new ecology. We really need healthy genes. We cannot allow natural selection to operate because, in the energetic equation of life, we would not have the resources to raise enough children. Besides, who wants to see their children die?

Sociobiology is the science that discusses reproductive strategy and one of the main points it makes is that due to the unavoidable high initial investment the female makes in reproduction and the overall limit on the potential number of children she may have, hers is always a strategy of quality where the male's strategy may be more one of quantity. The quality in this case means genes. That theory then says that this means the danger of birth defects is going to instinctively matter greatly to women and considering how conscious of genetic health many women can become during pregnancy, that seems true. Yes, men are concerned and will become more concerned as the understanding of genetic consequences becomes common. Also, monogamy places him under similar constraints as the women, but it is women's instincts that are going to drive artificial selection.

If we accumulate a growing genetic load, humanity will not be able to sustain civilization. We will revert to some kind of tribal, perhaps feudal ecology where natural selection again operates. This will be a time of disease, warfare, ignorance, hunger and short lives. Remember, our ecology is our life support system and civilization can support far more people than any other ecology. Historically, when civilizations have fallen, 95% of the population has disappeared. Some left, but many died. In this world, we are far more dependent on our civilization and the world is more crowded. There are not as many places to go to.

We must find a way to compensate for the great reduction in natural selection. That is the most important point of this discussion. The solution must be safe, practical, economical, and ethical.

4. How – Artificial Selection

Somehow, we have to compensate for the changes we have already made, especially the reduction in natural selection. The only effective way I can see solving this without causing overwhelming moral problems is to use pre-implantation artificial genetic selection to replace natural selection. The basic idea is that when a woman wants to have children, harvest 100 eggs from the female and fertilize them with her mate's sperm. Do a genetic analysis of the zygotes and pick the few with the least broken genes and the best combination of genetic potentials from both parents, that are implanted and brought to term. It may not even take the intrusive procedure of harvesting eggs from the woman. They can already use stem cells to produce sex cells from other tissues including skin, but that is not the issue. The initial technology of pre-implantation selection is not that difficult. It has been pioneered in a few countries including in the United Kingdom by Dr. Wells and in China by Dr. Wang.

In ways, it is little similar to current in-vitro fertilization techniques, because in that case multiple eggs are fertilized, but instead of genetic analysis, the doctor always selects the most vital and fastest growing embryos to implant. That can have a similar effect to artificial selection. Already limited genetic screening of known diseases such as cystic fibrosis has been done, but it does not have near the potential of true artificial genetic selection. The list of known genetic problems causing physical and psychological diseases and failings is very long and getting longer every day.

A key point of this is that the technology does not appear to be particularly expensive, especially when compared to a problem pregnancy. Also, the price is rapidly going down. The current cost is about $1000 for a "full genetic scan". Ultimately, I would expect the cost to be much closer to $10 and optical scanning will be even cheaper, but what is the value of removing the BROKA gene or many other genes with comparable effect? It will offer an amazing potential to everyone. This is important because everybody is going to need this. This is not just for the rich and elite like Sci-Fi always seems to portray. This is something that everyone who uses genes for reproduction is going to need to use. So it has to be economical.

In the news is the quite amazing gene-editing technology called CRISPR which is going to make some biotech firms just truckloads of money, but it is not the right tool for this job. Natural selection is a rather general effect. CRISPR is a much finer tool for producing a much finer effect. It can even create new genes which we currently do not need, but eventually may one day. Right now we need a larger effect than CRISPR can offer. It may eventually be usable to modify single point mutations, but the larger genetic changes caused by de novo mutations are not likely to be fixable by CRISPR. I describe it as that in the garden that is the human genome, CRISPR is like a weeder that can handle individual little problems. We need something much broader, perhaps like a lawnmower or herbicide. You can do that with artificial selection. You can handle a number of separate problems at the same time. You can handle larger gene changes that CRISPR cannot. Artificial selection is inherently less expensive because both techniques would use the same genetic analysis process, only CRISPR would require it multiple times as well as requiring the CRISPR process itself

There is another issue in that CRISPR can raise various safety and ethical issues that artificial selection does not raise. All genes husbanded by artificial selection have, so to speak, been vetted for safety and side effects by nature, which is very demanding.

At one point, research indicated that indeed the initial results from CRISPR were too good to be true. Apparently, it could cause "un-targeted" genetic changes to occur. Then further research indicated this was not the case. More recent research is showing an increased incidence of cancers associated with CRISPR and more un-targeted effects including ones that were not previously noticed. It will be a while before we really know in detail about what really happens when using that powerful technology. It may never be widely usable on humans. Again though, at a minimum, using CRISPR will lead to higher costs by requiring increased process and analysis.

Current genetic sequencing technology is based on making copies of the DNA using DNA polymerase.

While that technology is remarkably effective and continually coming down in price, it is possible that far cheaper optical screening by AI's (a task they seem good at) could be used as an initial screening step of comparing the genes of the zygote to the genes of the parents to look for breaks or major changes. Again this would offer a significant reduction in cost. In a way, this is how I originally envisioned the technology, before the revolution of modern sequencing techniques... and it really was an amazing revolution.

If that natural selection rate is around 50%, the actual "unfitness" rate will be lower. Natural selection does not just select against the unfit, it also selects against the unlucky. Using artificial genetic selection, we should easily be able to increase the selection rate to 90% or perhaps whatever is needed, desired, or economical and the element of luck would be removed so that compared to natural section, the selective effect would be far greater. That offers a lot of potentials. As critically important as it will be for humans to husband their genes using genetic technology, we are lucky that there seems to be a practical and economical technical solution available for the problem. Except in terms of hybridization, it is likely that artificial selection would not be needed for every generation, but it is fairly predictable that once something like this is started, it would become a custom.

5. Morality of Genetic Husbandry

It is time to pause here. Reproduction is one of the two foundations of ecology. It is the essence of human survival. It is also a foundation of morality and like everything related to human survival, it must be carefully considered in moral terms from start to finish. We need to use artificial selection, but this is a powerful technology with all the risks and dangers of powerful tools. To safely use it, it will have to be considered carefully in moral terms to try to avoid mistakes. Not only that, but as soon as you start talking about children, family, and human survival like this, there are ancient moral instincts that are far older than your intellect and that can actually be more powerful when it comes to decision making. Those moral instincts suddenly come to attention when the discussion goes in this direction and these are conservative instincts, just as life is the ultimate conservatism. Change for the sake of change is very dangerous. Those instincts have the final say in human activity and must be satisfied. So here must be an initial discussion of artificial selection in terms of morality.

Morality is considered in greater detail in my other writings. Generally, it is considered to be how we decide what is right and wrong. There is no general agreement as to how that is determined. Here, a moral system is defined in terms of biology as what would be called a "survival strategy" and has a great deal of associated moral instinct as well as education. It is the most basic meeting of Nature and Nurture. That definition is supported in detail elsewhere and does include consideration of moral systems from religions and philosophies as well as other places in history. Many different institutions husband and teach morality. Here though, the consideration of morality is kept to a minimum and a biological definition of morality being a "survival strategy" is used.

There is an assumption made here that follows that point, that all of this discussion is based on. That is that humans have a strong instinct to survive. Perhaps that might seem obvious, but it cannot be taken for granted. (We have a strong instinct to have sex, but the instinct to have children is not as strong.) Very early on, I examined this premise to get a feel for the nature of human survival instinct, and interestingly the instinct to survive seemed expressed as creative values. It is broader than just children and family, having other more subtle, but very important expressions as well that include artistry and moral justice. Human survival instinct is quite powerful and as basic as was expected, but perhaps more complicated. It cannot be taken for granted in this discussion though. This is discussed in detail elsewhere and is more important than one might expect.

* * *

In those moral terms, artificial selection is moral because it will lead to healthy children, healthy families, and healthy communities. These are the foundation of human survival and morality.

It also offers the potentials to allow us to adapt to this new ecology we must create.

There is more to it than that. Genetics is the most fundamental form of wealth there is. It is a wealth we can accumulate generation by generation and pass on to our descendants. It cannot be stolen or hoarded. It is never depleted except by mutation.

Also, there is a moral poetry to artificial genetic

selection. Genetics work additively. If you are healthy, intelligent, and attractive you may not think you need much improvement in your genes. You will, both because of natural deterioration over time and because we need to adapt a lot. Still, if you are gifted and perhaps have an IQ of 120 or so, artificial selection might be able to raise your descendant's IQ by 20 points in a few generations, which seems a good improvement. But what about a person who is less gifted and perhaps has an IQ of 80? In those same generations, it should be possible to raise their descendant's IQ by 40 points. Because of the additive nature of how genetics work, what you do not have is the easiest to add. That goes for beauty and health as well. A family with a history of an inherited health problem should be able to shed it easily. Those who have the least, have the most to gain from this. That is a nice thing in moral terms.

Keep in mind that while competition is a process of biology and evolution, what is more important actually, is being adapted to your environment. Do not mistake "survival of the fittest" as primarily being the winners of the reproductive competition (though it is more important in highly polygamous mammals), the "fittest" is the group in the species that is fit enough to survive and reproduce in their environment. That is rarely limited to some "alpha" or small group. That would lead to dangerous evolutionary bottlenecks. Unfortunately, humans just often think this way, but it is not suitable for the strategies that we have seemed to have wanted through history or will need in the future. We are in an evolutionary race that it takes many to win. We need to get past our Darwinian driven instinctive infatuation with the idea of an alpha.

What is the value of health? When it comes to the question of health or wealth, it is pretty much agreed that health is the most important or else that wealth cannot be enjoyed. If you want to be more objective and put a monetary value on it, health is of great value. There is often talk of lost days of productivity due to illness, but what about inherited weaknesses and disability? What is the cost of medical care to a country? Nations measure the cost of health care as a part of their entire gross national product, often in the area of above 15%. Ask a doctor and they will tell you that most of the expensive chronic diseases have genetic foundations. This is a technical society. What is the value of health and intelligence? What is the value of beauty in society? We do not talk about it a lot, but its value is great.

We do not generally judge nature or natural selection in moral terms, but if we did it would be known, as in the past, as red of tooth and claw. Judged by any human standard, natural selection would not be called moral. It would be called ignorant, brutish, and brutal. Artificial selection judged by that standard would be far more moral.

Then there is the other important moral consideration. It is that humans can not survive as more than animals without using artificial genetic selection. We would not be able to support civilization and would go back to being ruled by the brutal hand of natural selection where all energy had to be devoted to basic survival and reproduction. It would be a far cry from the moral goals that humanity has set for itself through history.

* * *

This is another discussion of morality related to artificial selection. It is a discussion limited in topic but be sure it is part of a very broad and sophisticated examination of human morality detailed elsewhere. Genetics is an important issue, but in many ways, it could be said that morality is a more critical topic to human survival. Do not think this is taken lightly or in isolation. This is one part of a far larger, carefully balanced consideration of human morality critical to human survival.

It seems a bit silly to write something like this without a direct reference to the elephant in the room. Some people clearly believe that disposing of a fertilized egg is killing a human being and so is morally wrong. Maybe that belief could be discarded as absurd and really, most people do not believe that. Their moral instincts do not warn them about it either. Still, no belief related to this topic should be discarded or ignored. It is about morality, so is just too important of a topic to not be considered. A moral balance must be struck, leaning towards the sanctity of human life or humans will not survive. There are no perfect laws and wishing there were is a rejection of the human responsibility and ability for judgment, as well as of our moral instincts. I have long examined this issue to come up with an appropriate guide in moral terms. This is not something I pull out of the air.

This is my best understanding of the moral views of smart, moral men and women that have articulated moral choices because they had to make them. Realize, in biological terms, the moral question would only be formed in terms of survival. In natural systems, it is not unusual for parents to kill children, including their own. It is the harsh calculus of natural selection that humans tend to morally reject. Also, artificial selection is going to dispose of many fertilized eggs for each child that is raised, even more than natural miscarriages do. I do not think that number should be considered a moral issue and I do not think the destruction of an embryo without a circulatory system is a moral issue. On that side of the moral balance are the survival of humanity and our greatest potentials. This discussion may one day apply to artificial wombs (a discussion of which, by a very smart lady, exists elsewhere. Unbelievably, that technology is actually starting.). Ultimately, I suspect we will settle on the commonest description of life as taken from the commonest Western moral foundation, the Bible, which very specifically defines human life as breathing. Certainly, artificial selection will not conflict with that or even the other more obscure reference mentioned, which was the presence of blood. I think I will leave the discussion here where it is for now. It need not go any further on that view for the purpose of examining the morality of artificial selection.

*   *   *

Another thing to mention briefly is that using artificial genetic selection could change other fundamental things about human habits. We compete for the fittest mate. It is part of what status is about. This is a huge and very basic aspect of human nature that existed and was important long before humans. It is part of very basic morality at both the learned and instinctive level. Genetic analysis and artificial selection could change factors related to this. Genetic analysis would almost certainly be a more effective way to get "good" genetic outcomes than the traditional forms of competition. It is not necessarily a change we want or do not want but it is a big change that should be recognized. That too will be considered more in other writings. Just as artificial selection will not remove natural selection, it is also unlikely to remove competition and that is a good thing. Though changing some current concepts of what status is would seem to make sense. Artificial selection can help us greatly, but it is only part of what is needed. Natural processes will and must continue to operate.

* * *

Based on moral instincts, almost all moral strategies and on practicality, pre-implantation artificial selection is not a moral problem. Obviously, some current religious thought opposes it but given the realities of it, that is not going to stop most people from using it and opinions change from generation to generation. By all reasonable standards, artificial selection is a moral strategy and the alternative is against every moral principle humans have ever followed.

6. What – Health, Beauty and Brains

Artificial selection is needed to replace the natural selection that has been removed, but it offers a lot more than that. We are going to need to use it to adapt to the new ecology and it will be used for that as well. If it is used to husband the great wealth that human genetics represents, what does that offer? Evolution is about gene frequency. We could use artificial genetic selection to increase the frequency of "good" genes, something that natural selection cannot do. Everyone could have excellent "health, beauty, and brains".

Nature does not guarantee that your children will inherit the genetic traits you respect the most about yourself or your mate. Somewhat the contrary. Artificial selection can offer that. Heredity is driven by chance. Artificial selection can be driven by human thought which can lead to a far better outcome.

Everyone can have a potential for the health, vitality, strength, endurance, agility, coordination, stamina, and speed of a natural athlete. Organs that do not fail, heart and lungs that do not weaken, and that provide circulation to keep all other organs and the mind healthy. We could all have excellent senses such as a finely discriminating sense of touch, vision of our world that did not fade or need correction, hearing that remains acute into old age, and a sense of smell that brings us the delights of nature and the taste of our foods. Smell has other potentials as well that few consider. Health is more than a physical thing. It is mental balance, endurance, resilience, flexibility, adaptability, and many other things.

This is not just about the health of youth. This is health that lasts a lifetime. There is a problem that humans are long-lived, but our organs and faculties age unevenly so that very often we have one weakness or another with age. We linger in life, mostly healthy but with one sickness or another that dominates our later years. This is because natural selection can mostly only operate until you reproduce. After that, nature is pretty much done with you. So we have a lot of genes that can fail with age. It is not a great mechanism for a species that has a uniquely long developmental period like humans and that must transmit its knowledge from generation to generation. Some people have genes that start failing shortly after reproduction causing heart problems, cancers, arthritis, mental deterioration, and other problems of age. Sometimes these even fail in youth. Artificial selection can be used to select for genes that do not fail prematurely. There will be trade-offs between efficiency, stability, performance, ability to hybridize, and other characteristics. There will be judgments we have to make when deciding what genes to husband and retain for our descendants, but those will be decisions that we can make and we can make them better than nature will.

Everyone could have great beauty, which would really confuse the whole issue, but I think we can deal with it. Really, we are more likely to have the beauty of health and vitality. Beauty carries a very high value right now. It is a visible manifestation of status. Also keep in mind that beauty, like health, is so much more than skin, hair, face, and form. It is how we walk, how we move, and it is the grace of movement. Beauty also includes your voice, how you talk, singing ability, and many other forms of beauty that we have no name for, but that do have genetic foundations. Like other traits, this is not just about the beauty of youth. This is about beauty that lasts through life.

And then there are Brains - intelligence is humanity's shtick. It is our special skill and will be even more important as our society and civilization develop. Intelligence is what got us here, intelligence is what will create the future and intelligence will be the most important ability in the future, though ultimately it will be in the service of morality and our instinct to survive. Intelligence largely developed as a social behavior, so it gives us our ability to understand ourselves and others. It provides your understanding of society and politics. It is your ability to communicate and to coordinate with others.

Intelligence is many things. Most importantly it is the neural net, a pattern recognition device. All our understanding comes from that. It is the source of insight. It is how we understand the world.

The mind has many specialized systems in it such as for language, one of our most powerful tools. Intelligence lets us use language not only for communication in so many amazing ways but it also represents a logical system that can be used for reason, analysis, and deduction. Intelligence is the foundation of what creates and uses culture. It is our intelligence that lets us use the cultural tools related to language that have given us so much more understanding. There seem to be intellectual abilities that could make it easy or at least far easier to learn new languages. All of these have genetic foundations.

An important function of our mind is spatial ability so that we can visualize our surroundings and world in our mind to find our way from place to place. We use that ability to map ourselves and even to mentally abstract ourselves in the world we inhabit. That abstraction may be a critical part of the genetic developments that seem to have started about 70,000 years ago and led to some of the unique features of what humanity is now.

Intelligence is critical to our usage of moral strategies. Our mind is our emotions, the loves, hates, passions, fears, happiness, and triumphs that are so much of what make us… us. It is the source of all our many artistic talents, the buildings we build as well as everything we design. It is the foundation of our ability to appreciate and understand art, something I think will become more important as time goes on.

Intelligence includes memory where all our knowledge is stored and organized. That is an extremely useful tool. We also have short-term memory that is like the RAM of a computer. We can only remember about seven things in our short-term memory. Enhancement of that would greatly increase our intellectual ability and it should not be hard to do because it already is different for different people.

There are numerous specialized parts of the mind, with surprising overlap. All of those different parts can be called upon by the neural net for analyzing the patterns it perceives. The neural net even has its own ability to use logic and reason for evaluating the patterns it generates. It can evaluate probabilities. Both abilities and others predate anything we consider language, and it may even have its own very primitive language. The neural net is our primary source of intelligence, problem solving, and understanding.

All of our ability to think has genetic foundations. Intelligence is wonderful. This world is getting more complicated. If you have ever tried to learn some new software or figure out a tax form, you probably would not mind being more intelligent.

Much of intelligence developed for creating and using tools. A basic description of humans in the past is a description of the tools they used. Our tools and technology are becoming more and more important to all our strategies. There are genetic-based predispositions that offer different skills for tool use and technology. We tend to think of genetic variation along the lines of tribe and race, but there is also great potential between various occupational castes. In the West - peasants, builders, warriors, scribes, priests, and the ruling caste all represent valuable genetic variation and potential that is almost never considered. We will need greater intelligence and more forms of it to create, maintain, and use the tools that our technology is offering us. We are making smarter machines and will need greater intelligence to use them, deal with them, and perhaps even to compete with them or learn from them. The foundations of intelligence are in the genes and mostly they can be added together. They are some of the greatest potentials that artificial selection offers. We will need greater intelligence to adapt to this new ecology we are creating. Intelligence must be considered to be a greater genetic wealth than even beauty.

It used to be that a person could get by with just good health and hard work (a moral thing). In the next ecology, I do not think it is going to be that way. In ways, the tribe was like an extended family and they took care of each other. A tribe had to have some smart leadership, but not all tribal members had to be smart. The tribal world had its challenges, but it was a simpler world of simpler society and technology. In a civilization where people are very different, a person is going to have to deal with the challenges of a more complicated world as well as humans trying to exploit them. I work to describe a society that is very mutually supportive, but that is an ideal. There may be less reason in the future for humans to exploit one another and machines may help us, but realistically life just usually involves competition. I have to think that it is not going to be a world of the lotus-eaters and if it is, it will not remain so. Our society is complicated by its nature and becomes more complicated and competitive just because of reproductive instincts. Humans are just going to have to be uniformly smart in the future as they have to have been uniformly healthy in the past.

7. The Three Levels of Artificial Selection

The next question would be, how can we safely and wisely take advantage of the developing genetic technology? First we have to make up for the problems of the changes we have already made, such as the reduction in natural selection. Then how can we best benefit from the great wealth that is the genetic diversity of all of humanity? Evolution is a change in gene frequency. We can greatly increase the frequency of good genes. That is to increase in the greatest wealth there is. It is a wealth that can not only be increased but it can also handed down from generation to generation. Because of the nature and requirements of civilization, genetic development also helps everyone's survival by helping provide the ecology they survive in.

This chapter is to describe the three levels of artificial genetic selection. Selection against the bad genes, selection for the good genes and selection for the stable hybrid. In other writings, I offer more discussion of how to safely and wisely use this powerful technology, but the most important point to keep in mind when considering genetic technology is a moral concept older than the pyramids of Egypt: Balance, known to the Ancient Egyptians as Ma'at. Balance is always an essential part of a strategy, especially when it comes to safety. Selecting for genetic extremes is just likely to be dangerous and unproductive. It is likely to be an indication of limited or short-term thinking. Keeping balance in mind will be the best way to avoid the dangers of any powerful technology and that especially applies to genetics. This topic is considered more in a discussion of morality.

The First Level of Selection Is Against the Bad Genes.

There are not a lot of bad genes. Natural selection does not tolerate them and got rid of most of them long ago, but each generation there will be newly broken genes (de novo) that need to be removed. In a sense, we are lucky, because while genetic technology is going to take some time to develop, what we need from it now will be the easiest part. Genetic examination and comparison of the genes of the zygote to the genes of the parents will show when there are breakages or significant changes. Any significant breakage or changes from parental genetic sequences are likely to indicate a de novo mutation that could lead to lifelong health problems, if the zygote did manage to develop to term. Beyond de novo mutations, current technology can already identify hundreds of genetic conditions that lead to major physical and mental health complications, such as the "BROKA" gene that can lead to several types of cancers. There are genes that similarly contribute to forms of mental illnesses. These can already be detected by genetic screening and there is no reason for afflicting our descendants with them. This can be done because we carry two sets of genes, but only pass one set of them on to our descendants. Usually, if there is a gene that has a weakness, the other gene of the pair does not have that weakness and that is the form to be selected for. If it is in both halves of the pair, then maybe it will have to wait until the next generation to remove it. Parents are going to be able to know about broken or problematic genes they carry long before they plan to have a family. They will know what their genetic weaknesses are. As children, we do not know. By the time we are adults we know about family problems of heart disease, arthritis, dental problems, mental health problems and other things that crop up in our families. We dread that they might effect us and may know early on that they are going to. It does not have to be that way. It may be that a genetic problem cannot be removed in a single generation. This is a long-term, ongoing project called life.

While there are not that many bad genes, there are many of those single point mutations. It would improve health over time if this were removed by replacing them with "normal" unbroken genetic sequences. It is hard to always say exactly what the effect of a single point mutation is going to be. A percentage certainly have no effect (10% would be a good estimate). Most can cause peculiar weaknesses in bones, organs, or nerves. They might not get noticed, except under certain conditions like during illness or after an injury. Some known serious genetic conditions, some fatal even, are caused by single base pair mutations though. At the same time, some might need to be watched rather than be removed, because they could be, might be, a beneficial mutation.

Both larger de novo mutations and single point mutations contribute to the rising rate of infertility.

There are viral genes that have been inserted in the human genome, some of which have become useful through evolutionary history, but many of which are of no use or can actually become activated occasionally and cause nasty diseases. The Human Genome Project estimated that 15% of human genes come from viral sources though the meaning of that is not exactly clear. The same is actually true of bacterial genes. They too have gotten inserted into the human genome, but little is currently known about that. For that matter, artificial selection does not just apply to the "human DNA", there is also hereditary material in the Mitochondria that are critical components in all cells. The exact same problem of genetic load is going to apply to that DNA as well, though there may be some differences in details. Being "older" in evolution, the effect should be slower. Mitochondrial DNA has been pre-natally clinically "replaced".

There are genes such as those that cause sickle cell anemia which have helped survival in some particular environments, but normally they are not a good thing. We will probably want to select against genes like that and instead solve the problem strategically like with mosquito netting, medical techniques, or other technology.

The adage of computer software must be applied to the software that is genes - "do not let perfect be the enemy of good enough". Perfect is not a word with real meaning and certainly not in life. Good genes are genes that work. We already have learned enough about human genetics to detect not just broken genes, but also some forms of genes that fail sooner over time or are simply weak forms of a trait. We will learn more over time and replace weaker genes with stronger ones.

The Second Level of Selection is for the Good Genes.

The second level of artificial selection is to select for the good genes. A lot of times that is just to select against inheriting the bad genes, but it also is about what you respect most about yourself or your mate. There is no guarantee that your children will inherit those features. Artificial selection will be able to change that natural 50-50 chance to 90-10 or higher depending on the desired result. You can choose to make sure your children inherit the best combination of genes for health, beauty, and brains that you and your mate have. Selecting to replace "bad" genes is selecting for good genes, but also you could select the best traits from each parent for your children to inherit. What do you feel are the best characteristics you or your mate have? Sometimes good genetics are obvious, but there is a lot to learn about what our genes offer us and there are going to be a lot of "superior" genes you have that you know nothing about or how they effect you. A good gene is one that is stable, functions well, and hybridizes well. This will be a long-term learning process. There are a lot of things we do not even know about yet that we will be able to select for. Humans have a lot of room for genetic improvement with the genes we have available. Using artificial selection could initially lead to great improvement, though the rate of improvement would slow with success. There is the term "golden child" to describe the rare child that seems unusually gifted, often from an otherwise very normal family. These gifted children can become more and more common. It will be a good day when artificial selection is only about selecting between two good genes.

Actually though, one must keep in mind closer, less lofty goals. So much of humanity is just struggling to get by even without having to face the leading edges of the great changes going on. Much of genetic improvement will just be about basic solid health, ensuring that the children inherit what healthy genes the parents have instead of leaving it to chance. A lot of times, selecting for good genes is simply a matter of just changing the 50% probability of inheritance of any gene and ensuring the inheritance of "good" traits that either parent has. In any generation, it will only one half of a gene pair, but the changes will accumulate. We must start to strengthen, to become healthier. Now we need to overcome weakness. Later we will achieve our great potentials for strength. Half of humanity has an IQ under 100. It should not be hard to make sure everyone has an IQ of at least that. That represents an incredible personal and social wealth. Just to raise the average IQ of humans by 10 points would produce an incredible economic wealth. We need to use the excellent genetic potentials that humanity already has before worrying about engineering new genes.

I have a friend that likes to remind me of the complexity of the problem of using artificial selection. Some traits exist on multiple loci, have linkages or crossovers. Current genetic knowledge shows that much of what makes us what we are is actually control mechanisms located in what has been called "junk DNA", meaning it is not well understood. A lot of it is about punctuation and how it effects other genes. This is referred to as Epigenetics. (It was obvious that it was not just junk.) Yes, this is going to be a challenging technology to understand and use. The thing is though that this is a system created by evolution and natural selection to be controlled and directed by natural selection, a very simple and brute force mechanism. Mostly all it can do is make a yes/no decision. Artificial selection can do a better, much finer job. We have a long time to work on improvement. We have already learned a fair amount about some of the simpler genetic systems that cause very human problems. The near-term problem is the broken genes. That will be a simpler problem, but we can already do more than that and our knowledge will grow rapidly.

Our genes are our greatest form of wealth and we have a great deal of that wealth from nature to work with. Using artificial selection, we can greatly multiply that wealth. We will need to in order to maintain the civilization we rely on for survival. We will eventually use it to become far more than we are now.

The Third Level of Selection Is For The Hybrid.

What is not talked about a lot is the hybrid. To students of humanity, it is well known that the coming together of different peoples is the source of most human cultural and genetic development. It has long been known that the tribes have always genetically and culturally mixed, but recent genetic studies on archaeological sites show that it occurred even more than was thought. It was not just travelers and migrations. Even more, it was slavery and warfare that led to the mixing of the tribes. When different tribes have come together, the offspring can combine the genetic potentials of both parental tribes. Unfortunately, it is not as simple as that, because the genes do not always fit together as well in the next generation and the statistical nature of genes means that some parental genes just naturally get lost.

Plant and animal husbandry professionals know about hybrids. When you combine two different strains or varieties of plants, the first generation has what is called hybrid vigor. It is healthier and more robust than the parents. It is much of the source of the high yield modern food crops that humans depend on for survival. They have great vigor from the characteristics of both parents, but the seeds of those plants are not replanted because the next generation is generally not as strong or healthy and their genes do not work or fit together perfectly after recombination. They are weaker than the first hybrid generation and sometimes weaker than either parent. Few offspring are lucky enough to have just the right genetic mix and combination such that those parental genes are going to fit together and work well. They are the hybrid that retains a combination of the parental abilities generation by generation. Over time, with selection and backcrossing, all the descendant offspring retain the potentials of both first-generation parents. This is the "stable hybrid" where the genetic potentials of the different parental tribes are consistently carried on generation after generation in each individual. It is sort of as if the genes of the parents were paint pots. The stable hybrid has both different paint pots from the parents and the paints have not mixed. (That is the best I could do.) While it can be costly in terms of natural selection and survival to reach that state, something that is mostly irrelevant to nature but very personally important to the individual, hybridization has been advantageous enough that nature has consistently selected for it. We are all descended from hybrids, as were they. It is simply how evolution works. The world belongs to hybrids and always will. Using artificial genetic selection with a far more developed knowledge of genetics, humans could take advantage of the benefits of hybridization while the natural disadvantages and problems could be minimized. The time of adjustment to becoming a stable hybrid could be cut down to a very few generations. Unhealthy hybrids could be rare and could be improved generation by generation using genetic analysis rather than the brutal culling that nature uses.

This book does talk about genetic differences between tribes. That has been a very controversial subject, especially because in the past science has been used to rationalize and support racism as well as justify race wars, but it is a discussion that must be had. To avoid those problems, this discussion recognizes and focuses on that the genetic variation of the tribes of humanity is its greatest wealth and its best hope for being able to adapt to the future. Using artificial genetic selection, the genetic drawbacks of mixing the tribes go away. Instead of looking at other races as being different or competitors, they must be looked at as having genes our descendants might want or need to survive in the future. Racism becomes an archaic artifact that is dangerous to us all. We need to look at ethnic differences, with the understanding that they represent the genetic wealth of humanity that will be required by all of us to adapt to the future. As it has been in the past, it is by hybridization that we will adapt to the future.

Another thing this offers is a different narrative. Many people look at races as static things that are fixed as part of the natural order and worry about how to preserve them to preserve that necessary social order and perhaps even follow some natural law. That is incorrect. Races are and always have been more fluid than that. Our short-term personal views just do not show that to us. Thousands of discrete races have disappeared, their genes absorbed into the genetic rivers that are the modern tribes of civilization. Yes, some have actually gone extinct. Some had few survivors. In most cases though, their genes have been distributed through the larger species and will survive as long as humanity does.

Egocentricity, the belief that one is special is an important component of a person's belief. It is part of our will to survive and compete. The problem is that it must exist in a balance or it can dangerously distort the person's understanding of reality. It can create divides or conflicts in the social group that is an important part of the human world and survival. Ethnocentrism can also be important in the same way. It is best if one thinks their tribe and people are special, but too much ethnocentrism could easily lead to a situation that is actually dangerous to survival. Partly that is because it is a type of old tribal behavior suited to a time long past, though clearly in tribal times peoples were not averse to some hybridization. They mixed. There are some isolated populations of the world now that clearly need an influx of new genetic variability for long-term survival. Any group practicing rigid racism will soon be surpassed in the changing world by hybrids. Also, certainly in the West, we use a fairly common morality, and racism conflicts with it, threatening the entire moral system. Conflicting moral systems are another thing. Races may mix more easily.

No one knows what the future will require of humans to survive, but I suspect it will be far different than most would think. Many peoples in the past thought they were superior and owned the world, as did the people that preceded them and the peoples that followed them. Not only does the future belong to the hybrid, but the present also belongs to the hybrid as well. We have great genetic wealth and it should offer a very bright future.

The latest tidbit of research is that in plants, hybridization can include polyploidy. I wonder if it can occur in humans. Nature is amazing.

About Hybridization and Filters.

When talking about hybridization, what probably comes to mind is races, which is why I commonly use the term. Race though is a bit deceptive because when talking about heredity, the tribe is going to be a far more accurate term. The individuals of a tribe are going to be relatively similar genetically compared to a race that may have incredible genetic variability. Consider what is called the Negro race. It includes the tallest and shortest peoples, the Watusi and Pygmy. Actually, that "race" includes more different tribes than all of the rest of the world. That is a lot of genetic variation. The same is true of all other groups called races, though there is also a valid meaning to the term as well. The members of a tribe though, typically have a similar physical and behavioral appearance. Race is a useful term mostly because people naturally identify race. That recognition is wired in and it is probably for more reasons than just our natural abilities to identify individuals. Recognition of tribes is wired in as well. Keep in mind that appearances are what evolve the fastest because it is external traits that deal with environmental variety. Metabolism and biochemistry are going to be far more similar between peoples than appearance and that includes psychological appearance which is as characteristic as physical features. That means that overall we are less different than we look.

What we need now to explain this is a bit of history according to the great British Geneticist C. D. Darlington. He described the sources of the agricultural civilizations that originated in The Fertile Crescent, The Indus River Valley, the Yellow River Valley, and Meso-America. He did not discuss African civilizations, possibly because without a yearly freeze to kill insects, traditional monoculture agriculture was not possible. Meso-America also had development problems because they had such limited availability of crops and domestic animals. The other three civilizations became what we describe today as Indian, Asian, and Western cultures. All of those cultures, commonly called races, are the hybrids of many tribes. Genetic and cultural progress in humans is mostly created by the coming together of peoples. Darlington described Western culture the most, presumably because that is what he had data on, but the pattern would be similar for all peoples and cultures. (It is intriguing but impossible to say what civilizations would have eventually developed in North America but as in other parts of the world, it was happening.)

According to Darlington (and this is now dated, but still seems basically correct), three tribes came together to form the Sumerian Civilization. They would have fulfilled the different needs of a civilization and the functioning of cities. (What made this more complex was the twin culture of Sumeria, Semitic Akkadian that rose to great prominence after Sumeria. They had to have mixed.) Peoples of the cities fulfilled basic occupations as castes including peasants, builders, warriors, scribes, and priests. Early Sumerian civilization was almost unique in Western history as a civilization not ruled by a military caste... yet. Priests led the society, directing the agriculture and public works. The different castes would have represented tribes. These would have been kept reproductively separate by custom and religion though hybridization would have naturally happened slowly over time. The occupations would not have hybridized, so the castes would have stayed discreet. Each occupational caste would have required different skills and temperaments, differences far more important than race. One of the most important differences would have been between peasant farmers and pastoralists with herds of goats and sheep. It is believed that farming really developed on the hillsides of the river valleys more than in the valleys. Beer and bread were common to the earliest Sumerian civilization, though it was not clear which came first. This pattern of cities spread and developed over thousands of years. Then the Semites led by Sargon the Great conquered Sumeria and added a military ruling caste. Always the military was descended from herding tribes because they were used to raiding each other's herds. This society was dynamic and spread civilization around the Mediterranean and even up the Atlantic Coast (where they met and mixed with neolithic hunters). Their skills of civilization would have been welcome wherever they went. They were the Minoans, Phoenicians, and other historic peoples. Over time, they naturally hybridized. Other tribes would have added to the civil society as their skills fit in. So while there would have been a hybridization between castes, especially due to war and slavery, the important hybridization would have been between different tribal sources that fulfilled the same occupational based caste. They needed to have the same skills and temperaments. They would have considered themselves to be of the same class and status, so they would have intermarried. The "tribe" of the caste would have evolved. Other castes would need different skills and temperaments. This continued for a couple of thousand years. Farming techniques, farmers, crops, and domesticated livestock naturally evolved and adapted as did the city folk. Technology developed and culture developed, sometimes internally, but often by absorption. Then the Indo-European horse herders from the Caucus areas came on the scene including the Greeks known as the Dorians and Ionians. The Romans and perhaps the Etruscans were Indo-Europeans. They kicked out the Semites as the military ruling class and may have originated anti-Semitism. Their civilization was even more dynamic and they followed the same colonization paths through the Mediterranean and up the Atlantic Coast building large cities that supported extensive trade. They "civilized" Europe, except for the Great European Forest that was "conquered" around the 10th Century. There is little written record, but the record clearly shows that the Celtics were a major genetic component that hybridized into this older group. It is often shown by the common strawberry complexion. It was the addition of this group that created the dynamic leading to modern civilization based on science and technology. There were so many other genetic sources as well.

There you have it. The modern Western world was composed of four basic tribal groups with many sub-groups and even stray tribes, all hybridized together over time. Only groups that hybridized with the core civil population originating in Sumeria could deal with the stresses of the city life as well as the diseases. It can be assumed that there were hundreds of tribes that joined this population, more or less successfully. Genetic analysis may one day tell. The occupations of the city stayed the same, though the occupational methods and practitioners had developed. The military-based ruling caste was international. The different nations of Western Civilization were ruled by a number of related families.

I do not know this kind of history as it relates to Asian and Indian civilizations. A scholar should be able to easily develop a similar ethnic history of those cultures. I have though been curious about the differences and may have some ideas about why they developed differently. Western Culture developed based on a hereditary, professional military ruling class that was "married" to greater and lesser degrees to the priestly ruling class. It was very stable in terms of lineage. Asian development was more fluid. "A peasant could become Emperor". As a guess, the early development of the repeating crossbow meant that armies would be composed of peasants in Asia. That disrupted the monopoly of military rule by a ruling caste that existed in the West. (It would be comparable to the development of weaponry during the American Civil War that ended the professional soldier of the Monarchy in the West.) In the Indus civilization, it is really hard to say for sure why civilization developed differently, sometimes it just seems they were more peaceful, but one thing stands out. The rivers that supported Indian Civilizations moved around significantly, sometimes hundreds of miles. That would have caused disruptions. Maybe it weakened a ruling class while requiring more cooperation among the civil populations. These are guesses. I could spend many years merrily studying Asian and Indian histories.

* * *

Back to hybridization. The previous paragraphs describe some of how hybridization has gone on in history and what the genetic materials there are to work with. The genetic variation is considered minor, but people are very different. Hybridization between "races" is one thing. Hybridization between castes is another. In ways, "races", the bigger visible differences between peoples are about separations in time, not adaptation (though skin color relates to the adaptation to UV light). Occupational castes describe differences between how people think and how we think is what is most important. Tribes or races that fulfilled the same occupational caste had to be fairly similar in nature and thinking no matter what the difference in their appearance. A farmer in Asia must deal with the same problems as a farmer in Europe. A herder in India must deal with the same problems as a herder in Africa. A scribe in any part of the world must deal with the same problems and those genetic potentials are what we will use in the future. These skills are what evolution has focused on the most since the time of the tribes because these skills were not as useful in the tribe ecologies as in civilization.

So variation in these important caste or occupational traits is greater than genetic statistics indicate. Where you might see more commonality is in hunting behavior. Oh yes, we do have the instincts for that, and you know it when hunting season starts.

Because of the place of war in history and the reproductive benefits the warriors had, the instincts of the warrior are relatively common. In the West particularly, there are individuals with well developed genetic-based "occupational instincts" of multiple or even all the civil castes. Think of an individual with the instincts of a builder and a scribe. They would be called a technician or engineer now. Regardless of our current occupation now though, we all had some ancestors that were warriors. War was so common, and warriors had such a reproductive advantage.

When talking about artificial selection and hybridization, it is probably best to consider physical and mental traits separately. You only have a couple of legs and one jaw. Artificial selection will be used initially to ensure that unbroken traits are inherited. As time goes on and we learn more, artificial selection can be to husband traits that do not break down with age or that are prone to cancer. Teeth are a good example where some families just have good teeth and some just do not. It is not about bad or broken genes even. It is just that some have better forms than others. Clearly, there will be trade-offs as well. Sometimes there might be a trade-off between performance and longevity. That is not what is critical in humans and everyone should be able to have excellent health, athletic ability, coordination, and even beauty of many forms, all of which should be long-lasting. Hopefully, when it comes to genes, humans learn to ignore trends.

Our mental characteristics are another thing. In discussions of Artificial Intelligence (AI) the point is brought up that evolution has produced a lot of shortcuts in the brain for problem solving, especially of problems we have had to solve in the past for survival. Call those shortcuts "filters" for now. They are like the specialized processing units used by computers for solving particular problems such as a GPU that computer Graphics Cards use. There are two things here to consider. The first is those specialized filters" and the second is... "filters" too. Unlike arms, the human brain can contain multitudes of these filters, composed of neuron patterns. Commonly they might even be called instincts because they are inherited behavioral potentials. They are genetically programmed ways to solve problems. Vision studies show that they can be pretty discrete. One neuron group may "see" vertical lines. Another may "see" horizontal lines. Another may "see" slightly off vertical lines. There are an awful lot of these neuron groups and they are very complicated. (An amazing study was done that traced the neurons to all the parts that made up nurturing behavior in a mouse.) All the data is collected together from these discrete parts and somehow finally ends up at the neural net for processing and then even as consciousness, a process still not really understood. Still, it is clear that there are millions, perhaps billions of discrete processing units that go into making up intellect. Also, unlike arms or teeth, they can be added in by hybridization. Humans primarily develop intelligence through evolution based on hybridization.

The thing is, it is not only about the "addition" of genetically based abilities and that is why I used the term "filters" to describe these genetic-based behaviors that range from instincts to nerve clusters to emotions. It is also because of what AI researchers found out and referred to as "filters". They found that they could make a machine that could solve a problem correctly perhaps 45% of the time, using a pseudo-intellectual process. They called one of these processes a "filter". They found that as they added filters (problem solving methods), the accuracy increased significantly to where with three or four filters the accuracy could be similar to a human's processing ability. With the advantages a machine can have, sometimes they can surpass humans at pattern recognition. It is assumed that they will continue to improve... but that is another issue. In humans, though it works similarly. The more ways a person has the genetic potential to solve a problem or recognize something, the more capable their intellect is. It is the basis of the neural net that is the pattern recognition function of the brain, our intelligence. Just like in a machine AI, the more genetic-based filters you have to add together, the more capable your pattern recognition ability is and the greater your potential for understanding is. The result is more than the sum of addition. Hybridization can add different filters together. Do not for a second forget how important training is, but the genes are important too and hybridization can add these filters in a genetic sense. These "filters" are what makes intelligence. Hybridization combines these filters in individuals, especially hybridization between castes.

As a little additional note, in my discussion elsewhere about intelligence, the question arises "can someone increase their intelligence"? Generally, the answer should be "no, learning is not going to help much", but there is another possibility. Making mistakes makes a person find new solutions to solving problems. They are adding ways to solve a problem. It is like adding software filters as opposed to the neurons that make up the hardwired filters in the brain. Training could be tailored to enhance this. A curriculum that leads students to make mistakes to learn from should be a normal part of education. The similarity though of using multiple "filters" shows why genetic hybridization can increase intelligence by adding genetic potentials for "filters".

Genetic Strategy

All discussion of strategy was supposed to be left out of this book to keep it simple. Strategy discussion was supposed to be in the Strategy book, but some of it crept in. That is OK though because this is about some strategy of Artificial Genetic Selection.

There is the old biological saw about asking if a committee of chimpanzees designing a human would choose to lose almost all their hair and have smaller teeth. What would current humans choose for the next model of humans? What traits and strategies would we need? This is meant to be just enough to describe some important points to get the idea across and offer important cautions. Again, I am focused on adaptation to getting us to the next ecology, civilization, which we need in order to be basically adapted to civilization in order to survive leaving our last ecology. Past that, I mostly leave the potentials for further human evolution to others who follow us. They should have more information to work with.

In a sense, all of this book about Artificial Selection is about strategy, but it is about replacing and imitating nature's strategy of Natural Selection. Natural selection does not require any judgment and is basically quite simple. Natural selection can only select against "unfit" genes, but humans are not going to stop there. They will effectively be able to select for good genes and even change the balance of traits in the population. So what should we do with the powerful tools of genetic technology we have besides using them to increase the "amount" of good genes we have available? Well, human nature, our physical nature, and behaviors, exist in a balance. Some people are short, some are tall. Some are slender, some are thick, some are aggressive, some are meek. Some are intelligent, some less so. Some are loving, some not as much. These all have genetic and environmental components. So if you are asking how might humans modify themselves in the future, you are asking how would the balance be changed in terms of genetics. Alert... this is risky stuff here, but it will be done and "needs" to be done to adapt to what is coming, so it must be carefully considered to minimize the risk. While this is about future human evolution, it can only be about individual actions. Only individuals have genes. Be careless with this and you risk the survival of your children. This is a case where it is appropriate to practice the conservatism of biology, that is to make the changes that are necessary to survival, but do not try for some ... form or another of greatness and certainly not some transient genetic fashion. That will be dangerous. Do not worry too much about unintended side effects though. They are going to happen until we are far smarter than we are now and life is about surviving surprises, adapting to them, and benefiting from them. This discussion does not include creating "engineered" genes, because no one has ever told me what they would create if they could. Besides, right now humanity has so much genetic potential to work with that we do not need any more genes to be able to adapt to the next ecology.

A great example to consider about this would be height. One of the most eloquent moral statements I have heard is "I want my child to grow taller than myself". It is not just a statement about height. It was a man's simple statement meaning "better than myself", in any way possible. Morality is sometimes simple and there is a lot of premium on tallness in our society. Any parent though, now given the ability to select for more height in their children, must ask is that a good thing or how much is too much of a good thing? For some parents that come from a typically short family, to suddenly have very tall children, well, is that going to work out good? Maybe, but without a balance of other proportions and physical systems, it may not. Can the individual's skeleton, muscles, and connective tissues support the greater strain of height? Maybe, maybe not. The change must be considered carefully. What is certain is that you can overdo it, so always work to keep a balance. Then you will be more likely to get a good outcome and avoid trouble. An extremely tall person can tell you how inconvenient it can be at times. Then there is what happens as you age. A large body becomes a physical burden. Gravity never ages. Selecting for extremes is almost always going to be a mistake.

That comment about height is a general trait, but occasionally we might want to change something more specific for a good reason. I guess I should mention that I have speculated that it would perhaps be practical if women were taller in general. Childbirth is very dangerous because of limitations on the size of the pelvis. If the pelvis was enlarged, the woman would have trouble walking. If she was taller though, then the pelvis could be larger and she could still move well. This is a very simple discussion of this possibility, but it is a good example of what we may consider, and it shows the perspective of the connectedness of any changes we might make. Besides, one-half of humanity tends to be much taller so the human body must be fairly adapted to that height and there might be other beneficial side effects as well.

I take a lot of this for granted because I study it, but I should not. The question is still often asked if our behaviors are from our genetics or are they learned. It seems obvious to me that it is going to be both, but that is not obvious to everyone and that is a pretty broad answer in any case. Konrad Lorentz gave the best description of this using the term "Behavioral Release". He said there were behaviors programmed into our genes, but they were not released until something in the environment called for them. He especially studied bonding behavior in birds, but the classic "fight or flight" response is probably the best-known case of this. Yes, you may even know how to fight or run, but until there is a threat to respond to, the instinctive behavior is not released and there is nothing quite like it. This applies to love, fear, anger, and other behaviors. When a programmed behavior comes out, it can come as quite an unfamiliar surprise, such as love or jealousy. You may never have experienced it before, but the behavior is there, waiting for an environmental release. It can be slow or sudden. An odd thing is that behaviors seem to even be able to be encoded in hormones. Hormones seem to be containers or even transmitters in the body of feelings such as anger, fear, love, libido, and other behaviors. This is what pregnant mothers may feel from their child. That phenomenon is not really understood currently.

What matters more than physical balance is mental and psychological balance. When you talk about changing humans, you are talking about changing human nature more than appearance. That is behavior. What would you change about that? We exist in a more complicated world technologically and socially than did our ancestors. How can we become better adapted to that? Well, we have a general term "intelligence" that covers most of it, but that is sort of more of a behavior than a nature. Intelligence facilitates drives, it is not one. What does facilitate the use of intelligence though is curiosity. That is one of the few balances I would change in humans if I could. That is, I think we should increase our general level of curiosity. It would increase our tendency to use intelligence and help adaptation to change. That covers part of it, including the technology part. What about the social part? We would naturally select for mental health. Our society is stressful and challenging, but to a large extent, that is selecting against weaknesses in mental health as well as making a workable society. What balances should we consider changing? Well, dominance and aggressiveness come to mind. Aggressiveness and violence was a pretty good strategy in the past, especially for the tribal world, but for many reasons including the real danger to the systems of civilization that we depend on to survive, violence is not as good of a strategy now. Dominance is a very Darwinian strategy and works in a polygamous situation, but when the long human developmental period requires monogamy, it is not as useful. Dominance instincts also seem to be the foundation for the human drive for power, a very ongoing problem for civilization especially when the drive for power is the foundation of greed. Dominance would be of less benefit to survival for humans using artificial selection and so not worth the risk. We need to shift that balance of our tendency to use violence, but, and I must emphasize this, the idea of losing aggressive potential seems very dangerous. Only the extremes in nature, the tendencies at the end of the bell curve to use aggression, should be curbed. This is talked about more in the Strategy book, but if you remove the potential for an aggressive response, you actually are increasing the risk of aggression because it may offer an opportunity for someone else to use the strategy. I could be wrong, but one of the closest things to a natural law I have found is that the only solution to violence is deterrence (at least now). Mostly, aggressive potential needs to be controlled by reducing the inclination to use it by education about the problems it causes and the benefits of the alternatives, but even more so by increasing the controlling mechanism of the potential, not reducing the potential itself. The controlling mechanism is self-control. Using that self-control has genetic roots but must also be taught. Only the extreme tendency towards using violence needs to be reduced. The inclination to use and the ability to use violence are two different things. Keep in mind that "aggressive" is also synonymous with "active", a very useful characteristic.

On the other side of this, the behaviors that facilitate social behavior needs to be enhanced. This would include cooperation and communication in its many forms, but there is another behavior and, as aggressiveness applies more to males, this applies more to females. We need to increase the balance of love. Now this is a general statement and the moral meaning of it is considered in the Strategy book, but there will be genetic parts to that as well. There is also the more familiar romantic "love" that is part of reproductive strategy. Sociobiology explains why love, the tendency to bond to a mate, is actually more of a masculine behavior (quite possibly derived from a mother's tendency to bond to her children). Yes, certainly women may express bonding as well because all behaviors are fluid between men and women, but many women do not have much mate bonding behavior. Love needs to be considered friendship, even between mates. It makes more sense and seems an important part of the description. That women are less likely to bond may seem counter-intuitive as monogamy is usually beneficial to them, but that may not require them bonding as is required of males. The males may do the bonding, they are the ones doing the most adapting to monogamy. When males bond, they give love and it is what they want, even more than sex. That desire for love has been a driver of civilization. Females do not always give love. It may also be that bonding by women requires a "behavioral release" that is less common these days. The lore is that if a man "saves" a woman (many possible meanings to that, it might mean "fights for"), she "is his" - she bonds to him - stimulus and behavioral release response. Yes, curious, but it is a common archetype as was pointed out to me by a woman. As in the Harry Potter story when Ron saved Hermione from the troll. (How else could Ron have ended up with her?) Females will bond to their children and may even neglect their bond to their mate if it is not strong enough. That is why a marriage should have friendship to start. Marriages based on friendship last longer because the female is bonded to the male before she bonds to here children and then that bond may last through the mother bonding to her children.

Moving forward, humans have a better chance of survival with a greater balance of love in many forms. Ultimately, love may be the key to the level of cooperation appropriate and necessary to the survival of civilization. Luckily most humans have the potential to be loving. If a group works together enough to form a team, it is based on friendship and love. Biology likes simple things so I am not sure if there is a difference between the two. It is not just important to teams. Its importance extends to the society. Then also think of how important it would be to any group in the confines of a spaceship for any extended time. In response to the violence of the Bronze and Iron Ages, the Zoroastrians developed a philosophy about love that was later taught by a Jewish teacher. Arguably, that philosophy led to an end of the insane conflicts of the Iron Age by offering an alternative to the Darwinian driven strategy of dominance and violence. It got a lot of traction and will probably be the key to future human survival. That belief was expressed as "love one another".

I will mention just a couple of other aspects of nature to consider. The first is "energy". The best way to understand this is to look at dog breeds. Some of the most interesting breeds are working dogs like a Siberian Husky but owning one without being able to provide an outlet for its natural energy is a problem for the dog and its owner. Humans vary greatly in their energy levels. Humans with a farming nature are able to put out great sustained energy like a working dog. Humans with a pastoralist nature are likely to run at a lower energy level but can immediately put out a high level of energy, like when a wolf shows up to attack their flock. Right now, civilization and industry have put a premium on the former, sustained output of energy. Farming types can work like machines, which has been very suitable for industry. In the future though, they will be competing with machines for that role. They could very well become like a working dog, without much to do to burn that energy. (There is reason to believe there would be useful outlets such as education in many forms.) This is another case where humans are going to need an interesting balance. Part of the problem of this is that energy level can control the usage of intelligence. Even great intelligence is of little use without the energy to use it and it does take energy. That is quite true for education as well. It takes energy. Too much energy though and you can burn out your heart or another system. (Then what if in 300 years we decide to build a starship and we need to change that balance again because even with the help of machines we need to work hard?)

Another possible genetic-based trait that seems worth mentioning would be "discipline", the ability and persistence to work now for what you want later. Even more, it is self-control. This seems so human and might be something that divides us from animals more than anything else. There are many different lists of virtues a person can practice. All have genetic foundations, but still need to be learned as part of one's moral education. Discipline though seems like the foundation of learning most things. It is about choosing what you want to be, perhaps what Maslow would have called "Self Actualization" where you react less and instead thoughtfully respond more by your own choice. In life, you tend to get what you work for. You may even have many material or genetic gifts, but you still have to work for them to really own them and to be sure that they do not own you. You have little, not even strengths or other virtues unless you have the discipline to work for, use, and husband them. That applies to material wealth as well.

So, humans have been under enormous survival pressures due to the changes that started with the end of the last ice age and the transition to farming and cities. What genetic balances have changed in that time? What can that tell us? There is the obvious physical toughness of the farmer and the mental resiliency and disease resistance of all the peoples that can survive in the confines and pressures of cities. There are indications that the slaughter of the World Wars reduced the aggressive nature of the European groups that fought in them. (There was the basic change in the nature of warfare in the West that started with the weapon developments during the American Civil War. The new weapons made older, traditional forms of aggressive strategy very dangerous to use in modern warfare.) Those aggressive traits came from Darwinian drives but now war has a large technical component that requires human strategies more than aggressive instinct.

The main genetic balance I would consider relates to our survival instincts. This is discussed more in the Strategy book, but it is important to mention here because not only are there genetic foundations to it, but because of its importance. If you asked me what evolution has changed the most in humans since we left the tribal ecology, I would say it has focused on our survival instincts. Now a very curious thing is that we have no common name for human survival instinct, and this is extremely important. "Animal survival instinct" is often mentioned, but instincts of any kind are rarely mentioned in relation to humans. This is probably because of the historical influence of the Catholic Church that has rejected the idea that humans have instincts because that would show that they are animals instead of divine. Well, we do have instincts, some of them quite powerful. As I said, survival instincts may have been the primary focus of evolution. Making a transition in ecology as large as we have, demands a resilience, adaptability, toughness, and a strong will to simply survive. Some animals are known to be very fragile and will weaken and die if taken out of their accustomed environment. Fragile does not generally describe humans. Where that balance will end up being is hard to say, but I see some challenges ahead, including philosophical challenges that are going to require a simple will to survive as well as a decision that humans may have to make that animals would not. That will to survive is the foundation of Why we make moral decisions about right and wrong as well as the decision to survive. It turns out that there has been great discussion about human survival instinct all through history. It is just that it is obscured. To develop conscious knowledge of one's own and other's survival instincts is an eye-opener and is a knowledge that can give great understanding and strength. Without its name though, it can be hard to recognize, understand or fully use. An individual must know the name of their survival instinct and understand it in themselves and others. That name is Faith and while religions have claimed that it is an unsupported belief in a God, it is really more important as an unsupported belief in oneself. It is like an emotion. It can grow slowly or flower suddenly. Like other emotions, you can see it in yourself and others. It is not a strategy of survival. It is a reason. It does not tell you what is right and wrong like moral instincts and systems. It drives you to choose between right and wrong. Faith is the name of the human survival instinct and is our most important instinct.

The key to using artificial selection safely is of course wisdom, but especially a respect for the wisdom of balance. There is a final trait to mention though that will be of great importance to human survival. If humans develop as I think they will, one of the most important traits for us to husband and learn will be Humility or we will cripple and possibly destroy ourselves with arrogance. They say that those that the Gods would destroy, first they make mad. The commonest form of madness is arrogance.

I met a couple of Jack Russell Terriers the other day while walking. They are a notoriously high energy dog breed. These two though had apparently been bred intentionally to be docile and low energy. Yes, they were, but when I offered my hand to them to smell, they ignored it. I have never seen a dog do that. A dog's identity is its nose. I wonder what else they lost besides their manic energy. We do not want humans to lose something that makes us what we are. The best idea is to use artificial selection to make humanity healthy and more of what we are and enhance our human potentials, perhaps even the annoying ones. We need to use strategy and education to deal with problems while avoiding using genetic solutions. We certainly are not smart enough to safely make important changes to our nature at this point. Especially not to carelessly lose something.

8. Morality of Humanity

It is time to go back to a discussion of morality because that has to be considered all through this, but this time in a broader sense. This time morality must be considered in the sense of humanity and the world. It is not just that the racial and ethnic and tribal genetic variability we have available is the genetic wealth of humanity, the greatest wealth there is. It is also that that variability is our only chance of adapting to the future. We need to adapt to very great changes. There is more to it than that even.

It is an old idea, humanity controlling humanity's genetic nature for reasons good or bad. It has scared humanity for as long as it has been considered and with good reason. Powerful tools demand great wisdom and responsibility to safely use. Also, we are nervous about it because it seems we are meddling with nature or that it conflicts with religious beliefs. Perhaps it is just because of the short-term view that is natural to humanity.

Well, right now we do not have any choice about using the technology, as the problem is already upon us. So it becomes a question of how to use genetic technology safely. If it is used to remove broken or bad genes, it will also be used to husband what humans consider good genes. The problem has always seemed to be that someone already has applied their ideology to what superior means, usually their race and people that look like they do. This discussion has gone to some length to prevent that. Our oldest moral knowledge from the time before the pyramids is of Ma'at telling about the importance of a balance that precludes ideology. Normally the nature of an ideology is that it is an unbalanced belief. The same knowledge of genetics that makes genetic technology possible will tell us that the diversity of human genetics is the wealth of humanity and that human survival has been based on the coming together of peoples. In order for humans to survive, husbanding our genetics wisely is only one of the problems that we will need to deal with using wisdom to find new and novel solutions. The solution to the problem of how to survive into the future is to use many solutions. It always has been. The human world has been changing for a long time, so get used to it.

*

Humans are amazing. This is not about concepts of superhumans or so-called Trans Humans. Those are dangerous dreams for the children. We need to be more than that. This is about the amazing potentials humans already have now for health, beauty, and intelligence that will allow us to adapt to a New Ecology where we can be comfortable and survive long enough to become what we have the potentials for now. Maybe after spending time there, we will learn enough to perhaps make informed choices about becoming something even more. For now, this is about finding a new ecology where we can survive. The old ecology and that way of life are gone unless we just want to be animals again. We must transition to the new ecology. We have all the genetic potentials we need for that. We have great genetic potential and can husband it to become great genetic wealth.

*

After the great disaster that was World War II, the victors looked around at the devastation and said "we cannot do this". Partly it was that they saw that it was as dangerous as the religious wars that had scarred so much of history. They knew of the wars fought to exhaustion, which began again when they regained their strength. That is how nature drives us. Now it will drive us to destruction if we let it. World War II and the one before though were a bit different from traditional European wars that were often family squabbles as much as for resources or religion. The World Wars reminded us of the indiscriminate mass slaughter of Iron Age warfare, something that had been deeply rejected morally as part of the rejection of Rome and the rise of Christianity. The stress of war causes the release and development of moral instincts in a person and after that time of global war, human moral instincts again said "no more". "We must not be consumed by war again". "We need to make a better world". When the war ended, it was also understood that it was not just about territorial conquest or resources. It had also largely been a race war in both the East and West. A moral decision was made that we were not going to do this anymore. We had to stop this ethnic warfare. We could not be having ethnic conflict and racial discrimination. It just is not going to lead to a future for humanity. It was a moral decision similar to decisions made before, but this time it was a decision that we were not going to judge people by race. Science was told to toe the line on this and not study heredity. That made sense because science had often been used to support racism. Now though, science needs to be looking at genes, genetic variation, and heredity, because we have a dangerous genetic problem we need to solve. Also, we need to understand ethnic variation, because to make it to the future, we need to understand the great wealth that is the racial and ethnic diversity available to humanity. It includes those "filters" that add together to make intelligence. It includes immunities and other physical traits. Genes are not mutually exclusive. They add together so inferior and superior have little meaning compared to the potentials of their combination. Genetic variation all needs to be husbanded. Jingoistic racism makes no sense and has no future. It endangers humanity. We do though need to know the different genetic potentials of all peoples that we can husband.

*

In the last century, an experiment was done with negative eugenics, probably with good intentions. It was basically replacing natural selection with a human decision, but the mechanism was functionally the same. The difference between that form of eugenics and natural selection though was that it was humans saying "you will not have children". That may have been less painful than natural selection, but it still was taking away the potential for a person's future. It was judged to be immoral because having children is the basis of survival and so the moral foundation of humanity. We said this is not moral. We are not going to do this. It was a moral decision.

In ways, artificial selection is similar to that in that the purpose is to improve the health of human and individual genomes. Artificial genetic selection is different though because it does not disenfranchise anyone. Instead, it is to give them a better chance of survival in the evolutionary sense by giving their children a better chance to survive. That is what moral is. Artificial genetic selection is about having healthy children that survive. It is not about not having children as traditional eugenics was about. Traditional negative Eugenics, like natural selection, can only select against individuals while artificial selection can select against bad genes and it can select for good genes. Selecting for good genes is a human strategy that nature cannot do.

Natural selection cannot select for good genes, only against bad ones. Natural selection tends to waste a lot of good genes, partly by being random and partly because it only works by competition. A lot of times, competition selects against quite fit individuals. That and the fact that natural selection involves so much chance makes it a really poor control mechanism, though clearly, it does function given enough time. In most cases, humans should be able to do far better at husbanding the genes than natural selection can. Still, it will be natural selection that operates on some of the most important things such as our ancient instincts.

*

I will make it clear here though that for all the agony of wars of conquest, religious wars, and race wars, we can exceed that horror by having demographic and population wars. That will be discussed more elsewhere but must be mentioned. While selecting for an instinct to have children just seems obvious, it absolutely must be kept balanced. Eventually, that too will have to be controlled by choice. It will most likely have to be controlled by custom and law.

*

We live in a rapidly changing world that is a challenge to adapt to. If you are gifted with health and intelligence, you may look at this future with anticipation and think that your descendants will not need to use artificial selection. They will, for genetic maintenance and to adapt, but what if you are not so gifted? The problem is that there are a lot of people in the world that are not so gifted, and they do not see a bright future for their children or themselves. The world is and always has been difficult. As was eloquently said, "the great mass of men live lives of quiet desperation". If you are not so gifted, you may be struggling now and wondering if there is going to be a place for your descendants in the future. This is not a good thing. A lot of people live with a fear of the future and a bit of despair. There are even some people working against the society and civilization because they know that the changes and what some call progress are not good for their children in any time that they can see. What artificial genetic selection is about though is that it offers them and their children a potential for a future. Generation by generation they can accumulate strong, healthy, superior genes, to when they can be well adapted to the new ecology we create and so they can be comfortable in it and compete in it. The morality of artificial genetic selection is that it offers hope to everyone. Nature would just kill them. We need this because it will be hard enough to create the new ecology we need in order to survive, so we need everyone to have a stake in the success of that future. No one is adapted to the future. Artificial selection offers the potential for a bright future, not just for a few, but to all individuals, all peoples, and to humanity. Nature would demand their death and replacement. Using artificial selection, they and their children can survive and develop. They can have a stake in the future. It is the only way for anyone to have a future. This can lead to a world for all and that is in accordance with human moral instinct and developments in moral strategy. Humanity has worked long to make a better world to live in for ourselves and our children. That desire is stated in the opening statement of the charter of the first modern democracy that so many nations have modeled themselves after: the Constitution of the United States of America, a truly amazing document. The Preamble to the American

Constitution is perhaps the greatest moral statement in any political document.

*

Decisions about the genetic nature of one's children must be left up to the parents. It is a moral decision and so among the most important that a person will ever make. Reason is a useful but limited tool for moral decisions. Moral decisions must be a free choice and is where natural selection will always apply. Maybe choice will not always be great, but it is the essence of free will and if there is a tradeoff, so be it. "Disinterested parties" have their own interests. I have actually wondered if there could be institutions in the future like religions, responsible only to people and moral law rather than governments and economics, that could help with this. Bad decisions will have to be judged by the society like a rite of passage.

We do live in a society though. One of the few laws I would suggest for regulating parental decision would be that a parent cannot choose to have their child inherit a broken gene that they did not inherit... (At least not without explaining the longer-term benefit to the child such as a trade-off or a potential for development.)

Genetics is life. It is biology, it belongs to nature and does not care about human truth. Biology holds truths we need to learn and pretty much need to follow to survive. Humans have often created truths and claimed they were Natural Laws, but they are not. Most "Natural Laws" of the past are now known as relics of history. We cannot apply our beliefs to nature. A human truth is unlikely to apply to nature. We are though designed to create our survival strategies, guided by moral instinct. Our beliefs will effect our destiny. We must understand nature to form our beliefs about survival. We cannot simply create ideologies that defy nature and then try to shape our nature to them. At best we will be unhappy, at worst they will result in disasters like what happened when Stalin caused the death of millions by applying the theories of heredity that fit his political ideology, to agriculture. It did not work no matter how many people died trying. Humans can craft strategies that nature cannot but we need to understand nature to do it. Nature cannot do what artificial selection can, but really it is meant as a carefully crafted imitation of nature to prevent disaster caused by changes we already made. That is a good way to be cautious. Nature has its limitations but ignoring nature or human nature without a great understanding of it and respect for it is going to lead to regrets.

Gene drive is a genetic technology that ensures that a gene is always inherited. Using gene drive in humans seems extremely risky because it would be meant to defeat some of the primary evolutionary survival mechanisms including adaptability and variability. Gene drive takes away choice and is only likely to occur as the result of an ideology. An ideology, by its nature, is not balanced, adaptive, or variable. It is not a survival strategy. (I have wondered if certain information, not beliefs, such as language and basic technology could be made accessible in our genes, but that is a strategic issue.)

Cloning skips a step in evolution. Perhaps it will happen, but even without the genetic downside, as a strategy of reproduction, it leads to a moral disaster for humans. We are a social animal, but cloning would conflict with that by removing the need for others and the compromises that involves. Those are part of what makes us human. Also, anyone that wanted to clone themselves almost certainly has an unrealistic view of themselves and does not believe there is room for improvement. That is a very dangerous belief and short-term, but in the extreme, cloning may be needed to solve a short-term problem. Cloning would prevent taking advantage of natural recombination as well as artificial selection. It could also perpetuate epigenetic changes from methylation which would not be good thing.

Avoid selection for physical extremes. It will end up badly. Even too ambitious selection for intelligence or some other desirable psychological trait holds danger related to adjustment and shows a lack of understanding of how finely adjusted the human mind is and the need for balance. Intelligence will need to be balanced with moral instinct and powerful survival instinct. If you ever are trying to achieve perfection, just give up right there. It means you are confused. Perfection relates to religion, not reality and history shows the great dangers of the pursuit of perfection. At the same time, aiming for perfection is not a bad definition of beauty and is a great way to achieve good quality. Aim for perfection in your endeavors and it will usually turn out good enough. Do not make perfection an ideology. It will be self destructive.

*

While on the topic of morality, what about the common moral argument of religion that our weaknesses strengthen us. It is quite true, but having genetic weaknesses that can cause a lifetime of misery and be passed on to our children does not seem the answer to that. Moral education is a completely different, critically important topic and probably better left to human design rather than the harsh, mindless vagaries of nature. If moral education is neglected, nature will step in and give a reminder. Moral education will become even more essential for a functioning society in the future, especially in a society where weaknesses are uncommon, but that is discussed elsewhere.

If genetic weaknesses are greatly reduced, will there not be greater discrimination and barrier to those that still have weaknesses, either genetic-based or from incidents in life? Yes and no. There will be more resources to help them, but yes, they will feel more alone. If they do have the moral strength and the help to overcome their physical or mental challenges though, their weaknesses will not need to be passed on to their children and their strengths can be. Again, this is really part of a larger moral argument and is written about elsewhere as strategy.

Your concerns with using this technology are certainly valid, but if you are not sure it is worth it, ask someone with genetic problems that run in their family. Ask someone that has had to develop the moral strength to make up for a weakness caused by their genes. Ask someone who has had to develop the moral strength needed to care for a disabled child and is terrified for that person's future when there will be no one that loves them enough to care for them. Yes, there are other moral considerations, but those relate to far more than genes and genetic technology and so are considered elsewhere. This is about survival based on avoiding the disaster of genetic load and on husbanding our genetic wealth. There are a lot of problems that must be solved. Do not get stuck on one of them.

*   *   *

The next chapter of this book was about Instincts. It became too long and seemed a digression so I removed it, but due to its importance I do think I should comment on it. It discussed our basic Survival and Moral instincts as well as Dominance, Cooperation, Status, Tribalism, and Love. It made a couple of useful points that I will summarize here. This is partly to describe some of our most important instincts but it is also to describe something about how they work. The thinking processes we are most conscious of are related to logic and reason. We can perceive and manipulate those thought processes. We can use cultural tools like language to hold them. Instincts are part of the more basic neural net that operate by pattern recognition. They are subtle, fleeting and evolutionarily a bit too old to be as clear as words, but they are important, useful and powerful. Instincts are based deeply in the genes. A person should know about them, something about how they work, how they can strengthen us, and how they can fool us.

Our instincts are old, especially our survival instinct that goes back to early in the history of life. The power of our survival instinct is amazing, but it tends to be hidden from us. This is partly because it is quiet, but also because in history the Catholic Church taught that humans have no instincts because that would mean we are animals rather than divine. That belief still exists as an important part of our culture. You are unlikely to even notice it though because it is just part of the cultural background. You need to know the name of your survival instinct to take the most advantage of it and it is your greatest strength. It is like an emotion though that can grow slowly or flower suddenly. You need to learn to recognize it in yourself and then you can recognize it in others. Its effect is so important and pervasive. It is one of the main answers or perhaps the main answer to the question "why". The problem is that it is culturally hidden by hiding its name. The name given to Human Survival Instinct is "Faith" and religions have claimed it is an unsupported belief in a God but more importantly it is really an unsupported belief in oneself and the value of your own survival. Like other instincts (based on neural nets), your survival instinct can be trained and strengthened by exercise, so maybe in the past it worked better as a part of religion. Now religion seems to lead to too many internal conflicts and focuses on other things like ideologies, so we need to recognize faith as a belief in the importance of ourselves, our society, our civilization and humanity. Morality is about how we choose what is right or wrong. Faith, our survival instinct, is why we make that choice. Why do you strive to do what is right? It is your instinctive struggle to survive. That survival instinct is known as faith. Look for faith in yourself and then you will see it in others. Look at what you and they do, and you will recognize when a decision to do the right thing was made. That decision was made because of faith, your survival instinct. Then you can consciously see the world in terms of our most basic and powerful instinct. You can see when decisions were made to do what is right in order to survive. You can see it in yourself and others. It is a power of life.

Morality is the outcome of our moral instincts and the moral systems we learn. We have instincts to seek and protect moral strategies. People fight over them endlessly. The main purpose of religions has been to teach and husband morality. It is what has given religion its great power (though sometimes it seems they teach hero worship rather than what the hero taught). Unfortunately, most of our moral systems have come to us through history with their authority based on precedence and the authority of a deity. Now, they are not adaptive enough for the changes we face and they cannot be defended. We need new foundations for morality based on reason and understanding or they will not be used. There is a problem with that as well though. Since before Aristotle wrote "Nicomachean Ethics", morality has also commonly been considered a topic in philosophy rather than just of religion, but that has always presented a problem for Western philosophy because philosophy is strongly based on logic and reason while often morality is not. Human thought processes, including intellect and instincts are based on neural nets that primarily operate by pattern recognition (though they can initiate logical analysis to verify the pattern they recognize). The pattern recognition of moral instincts may not use logic that is recognizable or accessible, so logical analysis of it often fails. Also, evolution is dominated by a very simple blind competition and so is very limited in what survival strategies it can create compared to what human thought can create. Our conscious biases may filter out its messages as unrealistic or undesirable. Our bias towards using logic and reason can make it hard to understand instincts based on intellectual processes in a neural net that evolved so long ago. Still, while not easy to understand and based on evolutionary patterns, moral instinct is a power that cannot be ignored. They can be very useful for protecting you as well as problem solving and providing understanding. That is what they were evolved for. You should get to know that ability, train it and exercise it just as you develop your intellect.

That leads to the first main point this is meant to describe. The human mind does not primarily work by logic and reason. It is a neural net or perhaps a collection of neural nets and so is a pattern recognition device. Though humans need to learn to use logic and reason consciously, the neural net can just automatically initiate logical processes to filter patterns it has recognized. Think of it in terms of vision, something of a specialized neural net in our brain that is sometimes called the visual cortex. You do not apply logic or reason to vision. Timing studies show the sequence of how vision works. Discrete elements of an image are seen. Then their location is determined, then an outline is formed. At some point, the pattern is recognized or not. You can reset the system by blinking, as when you are in a fog or the dark trying to recognize what you can barely see. Your vision is biased by genetics and learning. You see best what your ancestors saw and what you expect to see. You can prime the system to look for something (yes, logic can be applied). Most traffic on the optic nerve is actually from the brain to the eyes. If the neural net cannot identify a pattern, it often creates one. If a hunter is looking for a bird, they are likely to see many birds that are not really there but some clue triggers a recognition of the pattern of a bird. Instincts are the outcomes of neural nets. A bird does not think "last night was cold, I should migrate". Instead, many signals from the environment are recognized, combined, and evaluated over time until the neural net recognizes a pattern. Then a balance is shifted and the migration behavior starts. Instincts shift the balance between behaviors.

So be very careful trying to apply logic to instincts. At the same time, instincts are trainable. You can even talk to them though they cannot answer back. This can be useful if you are being distracted by an instinctive drive such as anger or love. They do listen. Here are a couple pertinent questions. Are the seven nerve plexuses that manage your limbs and organs part of your brain? Is the neural net that is called your visual cortex part of the larger neural net that you would call your mind that provides your intellect? How discrete is the older neural net for survival instinct or moral instincts from that neural net of your mind? These are maddening questions without discrete answers but considering them is necessary to understanding. Neurons work by "use it or lose it". If you were to just keep your eyes closed, other functions of your mind would start diverting neurons you use for vision to other purposes. So why is that important? For two reasons. The first is that considering instincts to be their own neural net, your instincts are trainable and get stronger with use. You need them to be strong. The second is that you can use them for problem solving. Your mind may be well trained to solve problems with knowledge, reason and logic but your greatest intellectual power comes from the pattern recognition ability that ultimately provides understanding. The neural nets specialized for survival instincts and moral instincts can also be brought to bear on what you would usually consider more intellectual problems. They have differences and can provide great problem solving power. That is one of the main reasons to train your survival and moral instincts.

That second main point this is meant to describe is something about moral strategies. Humans have two basic moral strategies or perhaps two sets of behaviors that make moral systems that correspond to two moral instincts we have. In a way, they are like the two concepts of genetic selection considered here. There is nature's strategy of natural selection that is blunt, brutal, inefficient, slow and cannot select for good genes, only against bad ones. There is a human strategy of artificial selection that is moral by human standards, efficient, faster and that can select for good genes. A larger lesson of my work is that often human strategies can work far better than the competition based strategies that nature is limited to. Consider two instinctive strategies that nature has given humans: dominance and cooperation. We know from fossils like "Lucy" that a few million years ago when humans embarked into the hunter-gatherer ecology that they walked upright but had small brains. Humans had great stereo vision from brachiating in trees and walking upright made that more useful. Also brachiating had given them great arm and hand development with dexterity that was enhanced by social grooming. We were not adapted to that new ecology though. We were just barely making it and we were popular prey for a number of predators including big cats. Survival at that time required cooperation and working together to survive as was shown by how much evolutionary pressure there was to develop the brain. That was the start of millions of years of rapid brain development, far more than any other land animal. It was for social behavior primarily but also served for tool use. It is from that long period of development that our current potentials for cooperation, communication, and gentleness come from. Violence in a tribe would endanger the whole tribe. A good example of that development would be Homo erectus. Then it seems that between about 70,000 years and 50,000 ago there was what might be called an evolutionary step. The real complexity of human evolutionary lineage is simpler than the Gordian knot, but not by so much. Indications are that the parietal lobe of the brain substantially developed, or it might be more accurate to say that the human brain re-organized. After that tools, society, art, funerals, technology, and other things human, greatly advanced. There are many theories about what was happening then that could cause such a seemingly distinct evolutionary development, such as a volcanic event or founders effect, but the reason is not clear yet. Humans changed though. We started killing all those big cats. Humans started killing everything. We dominated the environment and as happens when the environment is not the greatest challenge, other humans became the great challenge. This is when violence started to become a good strategy. Rather to their surprise, anthropologists have come to realize that homicide was pretty common in tribal humans. Remains of individuals that were in good enough condition to be studied such as "Ozzy the Iceman" very often showed not just wounds but old wounds that had healed. There are many indications of how common violence was and that it worked pretty well as a survival strategy. It could also give the males great reproductive success. There appears to be another genetic bottleneck in Europe, Asia and Africa about 7000 years ago that effected the Y chromosome. This was the time when agriculture was really taking off. The genetic distribution suggests that there were 17 women for each male. The best current explanation for this is patrilineal warfare where the males of some clans simply replaced the males of other clans. I wonder if "slavery" was involved but it looks more like a murderous masculine dominance behavior was an effective strategy at the time. This effect ended about 5000 years ago. Perhaps at that point, economics and culture had changed or only aggressive males had survived. (Zeng, T.C., Aw, A.J. & Feldman, M.W. May 25, 2018)

The value of cooperation and violence would change again with the start of farming and civilization. The challenge for farmers, like for early humans, was nature rather than other humans. This was also true for the other occupational castes of civilization. Builders did not benefit much from violence as a strategy. Neither did the scribes or priests. Civilization has always been about different peoples surviving together. Violence always worked against that. At the same time, there was a military caste that eventually became a ruling class that in the West lasted and developed into the Monarchy that still exists today. Violence was their best strategy for dominating civilization. It was a good reproductive strategy. This peaked with the Bronze and Iron Ages, times of extreme violence.

So if violence did not serve farmers, builders, scribes, or priests, who did it serve? Who became the warriors of society? These were the herding tribes. In the West, it was originally the Semitic goat and sheep herders that conquered the first agricultural civilizations of Sumeria and replaced the priestly ruling class with a military ruling class. Later it was the horse herders from Russia known as the Dorians, Ionians, Greeks, and Romans that replaced them. Herders made better warriors because they were used to raiding each other's herds as well as protecting their herds from wild predators. Importantly, they were somewhat aside from civilization. The Mongol horse herders were some of the greatest conquerors in history but had little use for cities or civilization. In the West, the conquerors became the ruling class and hybridized with the existing ruling class that was a mix of the priests, scribes and any previous ruling class. These were the spoils of war, as was power and there lies the problem. While civilization was developing based on creativity and cooperation, it was being controlled by groups seeking power and using the wealth of civilizations to wage war. Even before the Roman empire, the endless violence and slaughter had led to philosophies and religious beliefs adapted to coping with the violence and destruction. Christianity was one belief system that rejected the violence and, in many ways, replaced Rome that had limited strategies other than violence. That is not to say the violence or pursuit of power seemed to diminish much, but there was a change from "might makes right" being the commonest strategy to there being the moral alternative of "love one another". That was a human strategy.

That we have genetic potentials for both extreme cooperation and conflict is a bit amazing, but looking at history, it also seems necessary.

Just as humanity faces an existential threat due to genetic load, we face a threat just as overwhelming and dangerous in our natural Darwinian drive for mindless competition. We need to reduce that drive for domination and violence and instead use our instincts for cooperative strategies that we used for most of our evolutionary history. If there is anything like free will, it is here in the choice between following our Darwinian drives for blind competition that will destroy our future or following our other natural cooperative drives that can build an amazing one. Very few will try to claim that violence is the moral high ground anymore, yet we are immersed violence because of our history and those that fight for power. Unfortunately, it is a strategy that works pretty well and can mostly only be resisted by deterrence. The blind competition of different forms of war is exciting, but if not controlled, that strategy will inevitably destroy us. The only solution is to know about both of these instinctive strategies and make a human decision about which moral strategy we will use. Society must ensure that the powers of civilization are secured in the Charters and the Constitutions that are the laws of civilization. It must control and limit the power of individuals that would instinctively compete with society for power and can even threaten it. Society must understand and make it clear that it is not morally acceptable to mindlessly pursue power for ancient archaic instinctive drives. We rejected a military-based ruling class and will probably reject an economic royalty. Our society already accepts that a ruling class whose authority is based on violence goes against our material needs and our moral instincts. At the same time, some individuals now work on every level for the power to control the economic, moral, and intellectual creativity of civilization. This is a biological description of an ancient war between the rulers and those that they would rule but now is a critical time in that conflict just as it is for our genetic strategies. This is an ancient war that has never ended. This is the moral battle fought every day by all good men and women for freedom and for a future as more than animals. If we fail at either challenge, preserving our genes or preserving our civilization from those that instinctively seek to control it for their instinctive drive for power, it seems unlikely that humanity will have a future. This is all part of humanity's transition to a new human ecology.

The long and especially the short of this is that the same thing could be said about another human drive and that is "Status". So much of human motivation is about status which is based on the instinctive drive to reproduce with the most fit individual possible. If dominance instinct is about quantity and biologically is more of a masculine trait, status is about quality. In terms of sociobiology, it is more of a feminine trait because of the limits on reproduction by the female, so she competes for the fittest male often with beauty or by the result of male competition. Unfortunately, nature has made her instincts select success at blind competition as the judge of fit. When younger, women notoriously are drawn to "bad boys". (By the way, in humans there is endless overlap between what might be called masculine and feminine.) The thing is that in a situation where artificial selection is available, it could lead to a far better genetic outcome than an instinctive status strategy based on competition that is wasteful by design. There will still always be that drive for genetic wealth but instead of choosing the "last man standing", a person will need to consider other forms of wealth such as survival instinct, moral instinct, education, discipline, commitment, fortitude, curiosity, and other things, which are not going to be so simple.

Humans cannot survive alone. They are social animals that evolved living in groups known as tribes. There has been discussion about the likely size of these groups, but clearly it would often have changed based on the seasons as groups came together or separated based on the local and seasonal abundances. What a tribe was and how it operated would have been extremely variable, often even generation to generation as individuals and circumstances changed. It was a multi-generational thing though that perpetuated itself as life does. The important part here is that instincts evolved in individuals to support the tribe that was a critical part of their ecology and life support system. These included protecting the tribe and working with others for the benefit of the tribe. It could mean sacrifice for the tribe, even great sacrifice. In the tribal ecology, you usually just associated with others of your tribe. They were closely related and very like you in many ways. Your instincts and training were to protect the tribe. Observation shows that this could even mean putting the survival of the tribe over the survival of the individual. At the same time it also meant suspicion of others that were not part of the tribe and often a willingness to attack them for a number of reasons including just plain old blind evolutionary competition. Looking at the history of conflict in humans, clearly it was wise to be suspicious of other tribes (xenophobia). This was especially true in later human evolution when there were more humans encountering each other and they were more dominant in the ecology such that violence was a more useful strategy. Still, there were often benefits to working together and dangers to fighting. Deterrence was often the best strategy of survival. THings have changed. What matters is how these instincts and strategies effect us today.

So how does that play out today in the ecology of civilization where we encounter and work with people from other tribes that can be very different from ourselves? We have to be able to set aside most of our fear of other tribes. Our past is not gone but within a civilization we need to overcome our fear of others that are different and our blind instinct to compete with them so as to allow for the cooperation that makes a civilization work. On the other hand, we still need that instinctive and learned loyalty to tribe. One needs egocentricity to survive but it must be balanced with loyalty to the group as well. Now though the groups that one needs to be loyal to includes humanity and civilization. There will always be a balance between self interest driven by the our evolution and immediate needs. That applies to the individual and tribe. The trouble is that there will always naturally be what those that do not have that balance and think only of their own wants and needs. At the tribal level, that leads to racism, which is also intertwined with economics. In the long run, racism always leads to unstable situations that can endanger a civilization. There must be a balance between interests or there will be danger to the civilization. Civilization is now critical to survival to its individuals as the tribe was. The struggle to work together to make civilization work is a struggle for survival. It is familiar. It is the moral struggle for the larger things that humanity needs in order to survive as more than animals. It is not a war to be won. It is the daily moral struggle of all good women and men to personally thrive but also to protect their people, humanity and civilization that everyone's survival depends on immediately and in the future. We must understand our tribal instincts and choose what they mean to us, when they will serve us and when they will betray us. We need to understand that we are going to have to include humanity in what we believe is our tribe and treat it as such. We need to work to protect humanity and civilization.

Finally... love is a topic that is talked about quite a lot, though without a whole lot of consensus as to its meaning and importance. It is mentioned here in mostly terms of its instinctive nature. In ways it is as unusual as the extremely long and demanding developmental period of humans compared to other species. Monogamy is unusual in mammals.

In most mammals, the female raises the offspring without much help from the male and the male also uses the same resources which can make them a competitor. In harsh environments that demand it though, the male may stick around to help the female. (I am not sure of the reference to that.) In humans it seems like more than that. It can be a powerful bonding behavior and is primarily a masculine trait. Love is what men want out of a relationship even more than sex. In that sense it is counterintuitive, the male wants to be wanted. Even for instincts that seems an odd path. Of course, with the fluidity of all gender based behaviors, it can appear in either sex and when it does, it can be powerful. Love, in terms of bonding behavior to a mate, is more instinctive to men. For women, love of their children is more instinctive, but love of a mate is more of a choice. (This varies tribe to tribe.) A major determinate of whether humanity will survive long term is if women choose the importance of giving love. Without that, civilization will not develop the cooperation to endure. Any emotion can provide inspiration and motivation, but love is one of the main inspirations for the creativity that builds civilizations. Men will do anything for love including building a civilization, a good place for all. It may also be needed to replace the creativity that has been historically provided by war. This topic will be discussed more in the Strategy book, but it is an interesting, important and powerful instinct.

*

Morality and Instincts are extremely complicated and important topics that I will consider in detail elsewhere. Suffice to say that there are (at least) two things to keep in mind about it. The first is that it is your responsibility to set a moral example. The second lesson that must be understood is that though objectively speaking it is a choice between what instinctive strategy use, it is really the moral battle between good and evil, between civilization and the fall of darkness, between survival and extinction, between being an animal or something more. It is not a war that is won. It is the battles fought every day by all good men and women.

The most important moral lesson we must learn is that our destiny is in our hands and we must plan and act if we want to survive.

9. The Problem of Racism

This is a simple consideration of the dangers of racism. In biological terms, racism will prevent human survival. It is that simple. We have instincts and reasons that lead to racism, particularly fear of the unknown but also fears related to hybridization. The reasons are mostly inaccurate, and the rest will be removed by artificial selection, but the moral reasons are actually more important. This needs to be in the strategy section, but it is so important that it needs mention here in moral terms. Take Dr. King's warning literally: "We must learn to live together as brothers or perish together as fools". John Locke was one of the greatest contributors to the American Constitution including the right to freedom of religion. He said that because he knew there would be no peace and no future if there was ongoing religious conflict dividing the nation. To paraphrase that in terms of race he might have said: "until there is social justice and racial equality, there will be no peace". It is that simple, but it is not just about avoiding conflict. There are pragmatic, genetic reasons aside from conflict why racism will destroy any human future. No one knows what the future will be and it will be novel enough that we will need all of the genetic wealth of humanity to adapt to it. No one is adapted to it now. Progress has always been made by the coming together of different peoples and castes, adding their abilities. That is our only chance for survival. Going beyond the logic and reason laid out here, our deep, less visible moral nature that is part of being more than animals is going to demand social justice.

Never before have there been so many people in the world. Never before have quite so many different peoples come together in the world. It has led to increased awareness of competition. For humans in tribes, evolution and survival was always a team race. Now the team is bigger and more diverse but it is still a team race. No individual or tribe has the reproductive potential to win the race. We will win together or we will lose together. In a species like humans, there are different, separate populations but genetic analysis shows constant movement and flow. There has always been gene flow between tribes on the same continent. There has been some strong isolation from another such as the New World of the Americas and the Old world but that is the exception. Genetic analysis shows constant movement and mixing people anywhere migration or travel was possible. The past, the present, and the future belong to the hybrids, the peoples that have come together genetically and culturally. It is how humanity has progressed and will be more so in the future. Only the genetic wealth of humanity, represented by ethnic variation, can take us to the future. It is true that in nature, progress has also been by replacement, both individual and tribal. Morally and practically, that is unlikely to be as true in the future. The changes are too great. The winner now cannot be the greatest warrior. It needs to be the most adaptable and morally strongest. It will be the hybrid as it always has been. It can be all of us. We all have skin in the game and need to know that. As in the past, the adaptability we need will come from hybridization, but artificial selection will allow it to work far better without the high degree of wasteful culling and replacement that nature requires. Artificial selection will very efficiently allow the vast increase of genes for health, beauty, and brains of all peoples. Those potentials can be and will need to be preserved from all peoples. That will be important but for humans in the rapidly changing and demanding ecology we are entering, what may be as important though moving forward, is moral strength. That will be between difficult and impossible to select for artificially. It will be a choice between the individual and nature. We have no idea the sources of moral strength that we will find and be able to draw upon. That too will probably be a feature of the hybrid.

An important consideration here is that there is more important genetic variation than just ethnic variation. Caste variation is important as well. Using traditional ideas of superior and inferior in the context of race or caste is not going to work. Status is though how we naturally think and we need to think beyond that. Status is basically conscious individual genetic selection at the individual level. Artificial genetic selection will largely change the equation of that. Our instinctive drive for status has pluses and minuses in terms of social behavior but much of its genetic purpose can be replaced more effectively by artificial selection. We will still make conscious and instinctive decisions though about selection for our mate's mental and physical characteristics that will operate at different levels than artificial or natural selection operates. We do not know the future. We do not know what it will demand for survival. We do not know what the outcomes of hybridization will be, but we already know that the modern caste hybrid is much more capable than the more specialized castes of earlier civilizations. Hybrids just have more and different potentials. Those with the lowest status, historically often called inferior, may have talents such as those of the herders, that will be essential to the future. Hunters were at the top of the social strata during the time of the tribes, but hunting offers no status now. In most of human history, the military ruling class had undisputed top status but now they are replaced due to changes in technology. Most warriors have been replaced by technology as well and the remaining ones have had to adapt. It has happened before and will happen again. Our learned concepts of status need to adapt. We can see that our instincts about status are still strongly impressed by warriors. Only those that can adapt will survive nature's judgment.

It is not just about the risk of losing genetic variation or about social justice. It also about losing great cultural and technical wealth. It is about losing an incredibly productive part of our society. History is full of the cultural and technical creations of so many different races and cultures. Every culture has a different and unique art to offer humanity. It is not just that progress is lost to racism but also so many of the things that make life enjoyable. Yes, I am also thinking foods, but there is so much more.

Most claims of superiority are either by ethnicity, class, individual, or belief. Usually, it is about an individual or group and those that look like them. Nature does not care. Any ethnic supremacist is wrong by default because ultimately a hybrid is superior. That hybrid may have the best of their ethnic genetic potentials, plus potentials from another ethnic variation. The hybrid can also be more than the sum of the parts. Ultimately it is Nature that will always be the judge of individual and moral strength.

The criteria of what class is changes in many ways over time. Be careful of biases. We do not know what the future will require but consider this. For the past century, a primary class has been composed of people that can work like machines in factories and on farms. Now we face upheaval because machines have developed that can replace them more and more. This is like what happened to hunters and warriors before. Perhaps the most marginalized populations in the world now are the herders subsisting on the lands too barren for farming. What of them? It is true that their genes are perpetuated in all modern populations but is there any place for them in the future? That question is asked not for an answer, but instead to illustrate a point. The nature of a herder is to slowly wander after their flock, but always ready to become alert when the unexpected occurs. Well, does not that sound like what would be useful for someone watching over machines? They would ignore the boredom, but they would be able to be alert and ready to quickly respond to an unexpected event or problem. The point this was to make is that we do not know what the future will require, and our biases are not likely to inform us well. At the same time, much of what the future will be is up to us and our decisions. We need to choose wisely. That is what I will consider under the topic of strategy.

We are entering an unknown future, and no one

knows what we will need to adapt to it. Should we feel sorry though for those that are not adapted to the needs of the future? I have said that husbanding our genes will allow the less gifted to accumulate better and better genes for their children each generation until they have the ability to compete with those that seemed to start out more naturally gifted. Much of what they will need to succeed at that though is moral strength. It will be up to the individual's drive and moral commitment to use all their potentials and to make them what they will be. This is a critical and essential step on the path to human aspirations. A greater intelligence and strength will be needed to deal with the more complex world we live in and the strategies we will need to use in the future. Using artificial selection, we can preserve our genetic wealth to provide that part. Moral strength will be just as important though for using those other strengths.

If humanity is to go to space or the stars, it will take more capability than we have now. We will need to be far better at getting along. Only artificial selection will allow us to survive in and adapt to those places. That would be a strange ecology, requiring great technical skills, intelligence, physical ability, and social skills. If we meet aliens among the stars it will be because we chose to husband our genes and any aliens we meet will also have had to take control of their hereditary destiny. Nature cannot make more than animals.

Genetic Competition by Societies

Survival always involves competition and that will not go away any time soon. That is a good thing for many reasons including the individual challenge it offers us. Most importantly though, it drives improvement. I suspect that genetic husbandry will lead to an "arms race" between nations to improve the genes of their citizens. This will be driven by older dominance instincts and instincts for status. It will also be driven by human choice based on human understanding of the incredible moral, strategic, and economic value. It will be driven by disease as it always has been. America will have an interesting advantage there with our incredible wealth of different genetic sources. The thing is that instincts for dominance and status will not provide the answer to "why survive". Only human understanding and decision will do that and that will be driven by an older instinct, our most basic human survival instinct that is often referred to by the word "Faith". Those seeking dominance will try to manipulate human genetic nature for their own benefit and vision. Though they could cause a disaster in the short run, nature is not going to allow that in the long run. Survival will be a decision. Genetic development will lead further to where history has, to a kind of freedom of the individual that can only be ruled by consent, just as they will live because they have decided to. History shows that that individual will insist upon freedom in their society. We are all descended from ornery people and ultimately need to look out for ourselves as well as our society and species. Those who do not will see no reason to struggle for survival. Humanity has talked about equality and it is now enshrined in the laws of many nations. Beyond legal equality though, many people have great weaknesses. Artificial selection is not going to lead to some magical equality, but it can lead to everyone being smart, healthy, attractive, and strong. It will lead to a more realistic kind of equality for a people that are smart enough to make good moral decisions. Hopefully, they will also have good leaders available that are morally strong.

Our path will require increasing our potential for cooperation while mostly using strategy to reduce blind competition. In all cases, we must work to achieve balances, or we can be pretty sure of failure. Looking at anything from one point of view just does not lead to good results.

Because natural selection uniquely acts on individuals, it is hard to find any proof of what is called group selection in biology, but history shows that survival was not only about individuals. Tribes lived and died as well. Groups such as clans or families in the tribes survived and perished like individuals. Survival of a group was often based on leadership by a small part of the group. Survival has been about individuals or groups but now the change is so large that all of human variation will be required for humanity to adapt to the future. It is not a race to be won by an individual. It is a race that will be won by many individuals and groups. It must be a race won by humanity as a species. We will compete but on survival we must work together. When we embrace human genetic and behavioral strategies that overcome the blunt blindness of natural selection and our blind competitive drives for dominance, then we will be a new species: Humanity 4.0 perhaps. This will only happen if the moral struggle is fought each day by all good men and women to preserve a future for their children and humanity.

10. Wrap Up

The human world is rapidly changing, primarily due to the actions of humans. We have left the ecology we came from and need to adapt strategically and genetically to the new ecology of Civilization. This could be a long discussion about genetic strategy and technology but due to the removal of natural selection, increased mutations due to older parents, and other changes we have called human progress, we face an existential danger from an accumulation of broken genes. Artificial selection can allow us to ethically, and economically overcome that danger and at the same time allow us to genetically adapt genetically to civilization. Artificial selection is probably the only solution that will be available for a long time and it is already being done, at least in a simplistic form. CRISPR technologies will not do the job for the foreseeable future. Aside from being able to avoid the existential threat of genetic load, artificial selection can create the greatest wealth ever imagined by providing excellent health, beauty, and brains for every individual. It is moral because it will result in healthy children, families, and societies. Besides, it will be necessary for survival. Solving our genetic weaknesses will reveal other problems though, mostly coming from Darwinian driven competitive strategies such as dominance behavior and a drive for status. Those must be overcome by understanding and human decisions. That is what the next book on strategy is about. Luckily, those solutions are available from philosophies and religions, such as the greatest moral statement of all time: "love one another".

Again, this book is about genes within the context of this near term existential threat from genetic load. Solving that will allow the solution of how to genetically adapt to the next human ecology. Husbanding our genes will offer incredible wealth and potential for human development. I have to think though that there is another outcome from husbanding our genes that may be just as important. The related book to this that describes human survival strategies is meant to offer its own substantial value to human development and survival as well, but there is still something missing there too. That is the choice. We have great strategies from philosophy, religion, even politics and other sources. The Zoroastrian concepts of social justice and "to love one another" that was taught by Christianity are probably much of what we need strategically to survive into the future. The choice between the difficult task of working for a future for our children and humanity or the simpler evolutionary strategies of working for oneself represent the differences between humans surviving as humans or just as smart walking apes. Obviously the human strategies do exist but not everyone embraces them. So what could lead humans to make that choice? Philosophy, religion, history, etc. have only succeeded so much. What if there was an easier choice to start with that might lead to the harder ones? Well, I said that due to its universal need, artificial selection needed to be economical and ethical. It is an easy choice to make and women think about genes when they are pregnant. I suspect that making that decision to husband one's genes could lead to making the harder decisions about humanity's future, which in ways could be more important than even some of the benefits of genetic husbandry itself.

The key point of this is that the strategies of nature alone will not be enough for human survival in the future. Natural selection, with its huge reliance on chance, will not be enough to make us much more than walking apes. If we do not use human strategies of artificial selection to husband our genes, our society will collapse back to a time when natural selection manages our genes for us. It will be a brutal time of ignorance, red in tooth and claw. We will have failed to achieve the aspirations of our ancestors going back to the dawn of civilization and before. The same is true of our behavioral strategies. Instinct and environment will be enough to teach the win-lose strategies of blind Darwinian competition, but will not support the building and maintenance of a civilization. We need to carefully teach and use the win-win cooperative strategies created by religions and philosophies through the ages. If we do not know both strategies, balancing them individually and as a society, civilization will fail and we will simply be animals. If we use human strategies to husband our genes and learn the survival strategies to support civilization, we can have a future that is bright beyond our aspirations.

About the Book

I have been asked about references and citations, the traditional tokens of science. I have avoided these, preferring to emphasize older, more universal methods of proof from philosophy including reason and logic, upon a foundation of very basic science. (Maybe I am biased because I am good at logic or maybe because, as the philosophers said, you can depend more on logic for finding truth.) If you want to know more about that foundation in science, understand the basics of biology as taught by Charles Darwin, Gregory Mendel, Steven Jay Gould, E. O. Wilson, C.D. Darlington and so many others. I would say all of this could be found in a good high school biology book. Of course, the real understanding of that basic biology will be achieved through a great deal of hard work and thought. I spent a lot of time thinking about it, a lifetime really. Perhaps the hardest part of writing this was keeping it as accessible and understandable as possible. (I hope I succeeded but I know that understanding this requires time and a good deal of thought.) It would be easy to explain to a biologist, but it is something that profoundly effects everyone, so that is who it must communicate to. I have collected hundreds of articles to develop this understanding. I can provide them on DVD, ordered by year, if someone wants them, but the point was to avoid the need for that detailed knowledge or external technical citations to support this. It is supposed to all be here and accessible, with a bit of work and reason. I think there is one citation required about one detail though because it is not so commonly known and that is about the de novo mutations. The study of de novo mutations in "at risk" groups that is mentioned was found in the British Medical Journal Lancet Sept. 26, 2012. That information has developed greatly since then, but I have not pursued it further. That article was good to see. I waited for many years to see experimental verification of the breakdown of genes generation to generation, as it was the basic starting point of all my inquiries (after disease). Somebody could make a great thesis by accumulating the data on this problem from genetic analysis data from the many genetic studies done since then and interpreting what time frames it suggests for the progression of genetic load.

About the Author

I spent much of my youth in classrooms, the most interesting of which I found at the University of California Santa Cruz. I spent as much of the rest of my youth as I could at the ocean… What better place to learn the patterns of life? The rest of my time was spent in thought, trying to solve this problem. If you saw me, you would have thought me just another workman rather than a student. My jobs were always chosen to let my mind wander.

These books, this project, has been something of a labor of love or maybe of obsession. I have worked on this since my early teens. I have worked hard. I think, I hope, that I have created something valuable, a path to a good future for humanity where we can survive long-term and develop to become far more than we are now. I see great dangers, but I see even greater potentials.

This book describes our genetic toolbox we have to work with. Now let us see if I can finish the book on Strategy, part of which is strategies for how to best to use our genetic potentials and those tools of Artificial Selection to create a New Human Ecology.

www.ingramcontent.com/pod-product-compliance
Lightning Source LLC
Chambersburg PA
CBHW071433180526
45170CB00001B/328